Why Do We Watch Horror Movies

CRAFTED BY SKRIUWER

At **Skriuwer**, we're more than just a team—we're a global community of people who love books. In Frisian, "Skriuwer" means "writer," and that's at the heart of what we do: creating and sharing books with readers worldwide. Wherever you are in the world, **Skriuwer** is here to inspire learning.

Frisian is one of the oldest languages in Europe, closely related to English and Dutch, and is spoken by about **500,000 people** in the province of **Friesland** (Fryslân), located in the northern Netherlands. It's the second official language of the Netherlands, but like many minority languages, Frisian faces the challenge of survival in a modern, globalized world.

We're using the money we earn to promote the Frisian language.

For more information, contact : **kontakt@skriuwer.com** (www.skriuwer.com)

TABLE OF CONTENTS

CHAPTER 1: WHAT MAKES A HORROR MOVIE

- *Role of fear, tension, and surprise*
- *Use of setting, lighting, and music to create an unsettling mood*
- *Importance of a clear villain or threat*
- *Story structure that builds from ordinary life to terrifying events*

CHAPTER 2: WHY WE FEAR

- *Biological basis of fear and survival instincts*
- *Influence of past experiences and learned warnings*
- *Body and chemical reactions to scary stimuli*
- *Benefits and drawbacks of feeling fear in a safe environment*

CHAPTER 3: HOW THE BRAIN REACTS

- *Amygdala, prefrontal cortex, and fear responses*
- *Fight-or-flight reactions triggered by horror*
- *Influence of adrenaline, dopamine, and neurotransmitters*
- *Desensitization over time and the stress-relief cycle*

CHAPTER 4: OLD TRADITIONS OF SCARY STORIES

- *Roots in ancient myths and cautionary tales*
- *Oral storytelling around fires and local legends*
- *Festivals and customs built around spirits and ghosts*
- *Transition from folklore to written and eventually filmed horror*

CHAPTER 5: MONSTERS AND MYTHS

- *Origins of famous creatures in myths and folklore*
- *Symbolic function of monsters as deeper fears*
- *Evolution of classic beings like vampires, werewolves, and zombies*
- *Modern reinterpretations and cultural hybrids*

CHAPTER 6: HORROR GROWS IN FILMS

CHAPTER 7: GREAT DIRECTORS

CHAPTER 8: TYPES OF SCARY THEMES

CHAPTER 9: HORROR ACROSS THE WORLD

CHAPTER 10: THE PART OF MUSIC

CHAPTER 11: THE ROLE OF SPECIAL EFFECTS

- *Makeup, animatronics, and practical illusions*
- *Rise of CGI, blending digital and physical tricks*
- *Gore effects and careful use of on/off-screen violence*
- *Challenges and creativity in low-budget vs. big-budget setups*

CHAPTER 12: MORALS IN HORROR

- *Punishment for wrongdoing and cautionary lessons*
- *Characters facing guilt, redemption, or karmic ends*
- *Symbolic monsters representing moral or social warnings*
- *Complexities: not all horror is black and white*

CHAPTER 13: THE PEOPLE WHO LOVE HORROR

- *Variety of fans and their reasons for enjoying fear*
- *Collecting, conventions, and sense of community*
- *Emotional benefits: safe thrill and facing inner worries*
- *Bonding over group experiences and shared scares*

CHAPTER 14: FEAR IN DIFFERENT CULTURES

- *Local taboos, rituals, and beliefs about spirits or demons*
- *Communal vs. private expressions of fright*
- *Cultural roots of superstition and protective customs*
- *Intersection of old lore with modern life*

CHAPTER 15: HORROR ON SMALL SCREENS

- *Anthologies, TV series, and streaming innovations*
- *Shift from weekly episodes to binge-ready formats*
- *Balancing character depth with episodic scares*
- *Impact of wider budgets and relaxed content standards*

CHAPTER 16: LOW-BUDGET SCARY FILMS

- *Advantages of modest budgets: focus on suspense and creativity*
- *Found-footage phenomenon and viral indie hits*
- *Examples of classic low-cost successes*
- *DIY approaches, crowdfunding, and grassroots marketing*

CHAPTER 17: YOUNG VIEWERS

- *Children's and teens' fascination with spooky tales*
- *Need for age-appropriate content and guidance*
- *Effects on sleep, nightmares, and emotional health*
- *Balancing fun with caution to foster healthy interest*

CHAPTER 18: MODERN TRENDS

- *Social commentary in horror and "elevated" themes*
- *Streaming dominance and interactive storytelling*
- *Ongoing popularity of iconic monsters and franchises*
- *Blending horror with comedy, technology, and global styles*

CHAPTER 19: ENDURING POPULARITY

- *Timeless fears and cultural reinvention*
- *Financial success and cult followings*
- *Social bonding and community events*
- *Steady relevance through adaptation and fan devotion*

CHAPTER 20: FINAL THOUGHTS

- *Summary of horror's role in exploring human nature*
- *Lasting impact of monsters, myths, and shared scares*
- *Balancing tradition with innovation in the genre*
- *Looking ahead to future possibilities and ongoing fascination*

CHAPTER 1: WHAT MAKES A HORROR MOVIE

Horror movies are films that try to scare the people who watch them. They can show monsters, ghosts, or strange events. Sometimes they have loud sounds or sudden surprises. Other times, they rely on slow, dark scenes where something feels wrong. The key idea is to create a sense of fear or tension. People who enjoy horror often say they like the thrill. They get a rush of energy when they are scared in a safe place, like a movie theater or their living room.

One big part of a horror movie is the mood it creates. A good horror movie usually has lighting that is dim or filled with shadows. The sets might be old houses, lonely roads, or eerie forests. These settings make the audience feel uneasy. Filmmakers also use music and sound effects to raise tension. A quiet moment can suddenly become loud, and this can make a person jump. This surprise is often called a "jump scare." Even if we know it might happen, it can still catch us off guard.

Actors in horror films play a big role in making us feel afraid. Their facial expressions show terror, confusion, or shock. Their screams or cries can make us think something terrible is happening. The way they talk, move, or hide can also add to the tension. If a character looks scared and tries to escape something, we might feel nervous too. We imagine what might happen if we were in that position. This is part of the reason horror movies can feel so gripping. We see the world through the eyes of the characters and share their fears for a short time.

Many horror films also have a clear villain or threat. This could be a monster, ghost, demon, or even a human being who does bad things. The threat might look scary, or it might stay hidden until later in the story. In some movies, we only see a shadow or shape at first. This builds our curiosity. We wonder what the creature or villain looks like. Our mind fills in the gaps with our own fears. This can be even scarier than if we see the threat too soon.

Another important part of a horror movie is the story. Many horror films begin with normal life. The main characters might be friends or family. They have normal concerns, like school, work, or relationships. Then something strange happens. It might be an odd noise at night, or a door closing on its own. As the story goes on, the strange events grow worse. Eventually, the characters have to face the source of the horror. This structure keeps viewers interested because they want to see how it ends. Will the threat be defeated? Will the characters escape? Will there be a twist ending?

Some horror movies also deal with themes beyond just scaring us. They might talk about deeper ideas. For example, a film might show how guilt can haunt a person. Another film might show how fear can lead people to make bad choices. Some horror movies talk about topics like grief, loneliness, or anger. They use scary events as a way to show how these feelings can harm people or bring them together. Even though fear is the main draw, horror can also teach us lessons about life.

Horror movies can be full of visual effects that make the threat look real. Special makeup can turn actors into monsters or zombies. Fake blood, strange costumes, and odd camera angles can make us believe the events on screen. In older films, these effects were simple. Filmmakers used rubber masks, paint, and puppets. People could sometimes tell they were fake, but it was still fun. Modern movies use computers to show ghosts, monsters, and other scary scenes more realistically. Either way, the goal is to give viewers an intense emotional reaction.

Throughout film history, many scary ideas have been tried. Early horror movies focused on monsters like vampires or werewolves. Later, we saw stories about haunted houses or possessed dolls. Over time, horror filmmakers came up with new ways to frighten people. Some movies use a lot of gore. Others focus on a slow build-up of fear. Some make us jump out of our seats with sudden surprises. A variety of themes and approaches exist, and each viewer has their own tastes. Some like the classics with famous monsters, while others enjoy modern ghost stories or psychological horror.

Horror movies can also have humor. Sometimes, a film will include funny moments to ease the tension. This might involve a silly character who says odd things. Or it could be a comedic scene where the characters laugh at their own fear. This can make the scary parts feel stronger by contrast. When we laugh, we relax a bit. Then, when the horror returns, it hits us harder. It is a balance between dread and relief.

Many directors become known for their horror work. They might be recognized for a certain style, like using heavy shadows or focusing on a specific kind of story. Some prefer making stories about the supernatural, such as ghosts or demons. Others prefer realistic terror, like a dangerous person who hunts people. These directors learn to use lighting, camera work, and sound in ways that build strong tension. They also choose actors who can show fear in a convincing way. Over time, they might create a series of well-known films that horror fans enjoy.

Horror movies often reflect the fears of the time period they are made in. For example, older films might show anxiety about wars or diseases. A more modern film might show fear of technology or the unknown future. Horror can hold up a mirror to society's biggest worries. That is why sometimes we see a surge in specific horror themes. If people are afraid of a pandemic, you might see more films about deadly viruses or zombies. If people worry about advanced machines, you might see films about dangerous robots or computers that harm humans.

The way horror movies make us feel can be exciting. We know that the scenes are not real. We are in a safe space, but we experience strong emotions. We might scream or close our eyes. Our heart might race. But once the movie ends, we can calm down and talk about it with friends. This controlled fear can be a fun break from daily life. It is not for everyone, though. Some people get nightmares or feel uneasy for days after watching a scary film. Others find it thrilling and want to watch more.

The types of horror are diverse. Some films show large creatures that destroy cities. Others focus on ghosts haunting a house. Some center on a murderer who targets people one by one. There are also psychological horrors that play with the mind. They make the audience question what is real. Each type can scare us in a different way. Sometimes, it depends on what a viewer personally finds disturbing. A person who fears ghosts might find supernatural movies far scarier than a person who does not believe in them. A person who fears the dark might get tense watching scenes with limited light.

One thing that can make a horror movie stand out is the film's atmosphere. The atmosphere is the overall feeling or mood. Filmmakers create atmosphere by combining sets, costumes, music, and sound effects in a certain way. For example, if a film is set in a broken-down hospital at night, with flickering lights and distant screams, it can feel disturbing. Even if no monster is on screen, the sense of unease is strong. This is atmosphere at work. Sometimes, the suggestion of danger is scarier than showing the danger right away.

Horror movies also rely on plot twists or shock reveals. At the end of a movie, we might learn that the main character was the villain all along. Or we might see that the threat is still alive when we thought it was destroyed. These surprises leave the audience with a final burst of fear or uncertainty. Some viewers love these twists, while others might feel tricked. But it is a common tool in horror. Filmmakers want us to leave the theater still thinking about what we saw.

In many horror films, characters make decisions that might seem unwise. They might go into the dark basement alone, even though they suspect something is there. They might split up from the group when they know a threat is near. These actions can be frustrating to watch, but they also increase tension. They put the characters in danger, and viewers wonder what will happen next. It can make us shout at the screen, telling them to run or hide. This shared feeling of wanting to protect the characters or see the outcome is part of what makes horror engaging.

Some horror movies explore the unknown. For instance, they might deal with old curses or mysterious objects. We might see a book that summons ghosts or a mirror that shows strange faces. These items bring a sense of wonder mixed with dread. People ask themselves, "Could this actually happen?" Even if it seems unlikely, the movie makes it feel real. The unknown can be powerful in horror because it taps into basic human fears of what we cannot understand.

A horror film can vary in style depending on the culture it comes from. For instance, some cultures have tales of spirits that protect or harm people. Others have stories of ancient creatures that punish the guilty. These stories influence local filmmakers, and they bring these myths to the big screen. In one place, you might see ghost stories involving old shrines and silent apparitions. In another place, you might see giant creatures rising from the sea. Each culture's myths shape its style of horror films.

Horror movies also have different ways of showing violence. Some are very graphic, with lots of blood or strong imagery. Others are more suggestive, leaving much to the audience's imagination. The choice depends on the director's vision and what they think will scare people the most. Some viewers enjoy the shock of graphic scenes, while others prefer subtle hints. Both approaches can work if done well.

For children or younger viewers, many horror movies can be too intense. The images, sounds, and themes might be frightening. That is why certain films are rated for adults only. However, lighter versions of scary stories exist, such as animated tales about spooky events. They might have ghosts or monsters but keep the tone softer. This way, younger viewers can see small bits of spooky fun without being overwhelmed.

There is also a big market for horror merchandise. When a scary film becomes popular, fans might buy masks, posters, or collectibles of their favorite monster. They enjoy having items that remind them of the film. Some people dress up as these monsters for costume events, showing how they connect with these creepy characters. This is interesting because it shows that even though the monster is meant to scare us, it can also become something people like to see represented in toys or figures.

It can be surprising how many people are attracted to horror. Even though the films scare them, they continue to watch. The feeling of fear can be thrilling when it is safely on a screen. People often talk about the excitement of watching a horror film in a group. They jump at the same scary parts and then laugh about it afterward. This shared experience can create strong memories. Horror movies are also a way for people to face their own fears in a fictional environment.

Horror can connect to real-life themes. A film might show a family dealing with a haunting, but it also shows how they handle conflicts with each other. Another movie might show a community fighting an evil force, but it also shows the problems between neighbors. The scary parts might be over the top, but the human emotions can feel real. This mix of fantasy and reality helps keep us interested. We can relate to some of the characters and their struggles, even if the situation is strange or unrealistic.

Sometimes, horror films include moral lessons. They show characters who act in cruel ways, and as a result, they face a horrible fate. They might warn viewers not to trust strangers or not to disturb certain places. In this sense, they are like old cautionary tales told around campfires. The difference is that movies can use modern effects, music, and editing to make the story more intense.

Horror also gives filmmakers a chance to be creative. They can think of new creatures or settings that have not been shown on screen before. They can experiment with different camera techniques. They can use sounds that are rare in other genres. Because horror is meant to create strong emotions, it often pushes the limits of imagination. This can lead to very unique films that stand out.

In some horror movies, the threat remains unexplained. We never learn where the monster came from or why the ghost haunts that house. This mystery can be as frightening as any jump scare. When there is no clear reason, the threat feels bigger and more uncontrollable. Some viewers find this unsettling, while others feel frustrated by the lack of answers. It depends on personal taste.

Horror is also connected to the history of storytelling. Long before modern films, people told scary tales around fires. They spoke of creatures in the woods or spirits that roamed abandoned places. These tales warned people to be careful or explained things they did not understand. Today's horror movies are a continuation of that tradition. They just use the medium of film instead of spoken tales. Even though technology has changed, people's fascination with the unknown remains.

To sum up, horror movies are defined by the fear they create. Filmmakers use lighting, sound, acting, and story to build a sense of dread. They may show monsters, ghosts, or killers. They may focus on the unknown or on the dangers of human nature. Horror can teach us lessons, reflect our worries, or just provide a safe scare. Different people enjoy different types of horror, from old classics with rubber masks to modern masterpieces with digital effects. The common thread is that these movies aim to unsettle us, tapping into our deepest fears in a way that can be both thrilling and meaningful.

CHAPTER 2: WHY WE FEAR

Fear is a natural feeling that helps keep us safe. When we sense danger, our body reacts. Our heart might pound, our palms might sweat, and our mind might sharpen. This reaction can help us run from threats or stay alert. In ancient times, humans needed fear to survive against wild animals and other dangers. Even though modern life is not as risky for many people, we still have fear. Horror movies play with this instinct. They place us in scary situations on screen, and our body responds in a similar way, even if we know it is not real.

One reason we fear is because of the unknown. When we do not understand a strange noise at night, we might think it is something bad. Our mind fills in the gaps with the worst possibility. We do this because it can keep us safe. If we assume the noise is harmless, we might ignore it and end up in danger. But if we assume it might be a threat, we stay alert. Horror movies use darkness, unclear shapes, or offscreen sounds to trigger this fear of the unknown. We see a shadow move across the wall, but we do not see the source. Our mind imagines it could be anything.

Another reason we fear is based on past experiences. If we had a bad encounter with dogs as a child, we might fear dogs as adults. If we got lost in a dark room once, we might dislike darkness now. Horror movies can remind us of these experiences. A film featuring a scary dog or a locked room might bring back that memory. Our heart rate goes up because we recall the panic we felt before. This is how personal experience shapes what we find frightening. Not everyone fears the same things. One person might be terrified by spiders, while another might find them interesting.

Fear also comes from how we learn. Some of us hear stories about certain dangers from parents or friends. They might say, "Do not go out at night," or "Strange things happen in abandoned places." These warnings can stick with us. When a horror movie shows those situations, it taps into our learned worries. We might feel uneasy because we have been told those places or actions are risky. Even if we know it is fiction, that sense of caution is there.

Many psychologists say fear can be both helpful and harmful. It is helpful when it protects us from real danger. It is harmful when it stops us from doing normal things. For instance, if a person has a strong fear of heights, they might avoid going to high floors even when it is safe. Horror movies play with the line between helpful and harmful fear. They give us a controlled scare, letting us feel fear in a safe environment. This can be exciting for some people, but too stressful for others.

Our body's fear response involves chemicals like adrenaline. Adrenaline makes our heart beat faster, and we might breathe faster too. Our muscles tense up, ready to act. This is the body's way of dealing with threats. In a horror movie, we might feel a smaller version of this response. Our heart races, we grip the armrest of our chair, and we feel tense. Once the scene ends or the movie finishes, we realize we were not in real danger. Some viewers find this sensation thrilling. They like the rush of adrenaline because it feels like a roller coaster ride.

People also fear things that are against nature. For example, we expect the dead to stay still. A walking corpse breaks that rule. Seeing that in a horror film can be unsettling. Similarly, we expect dolls to be harmless. A doll that moves or talks on its own feels wrong. This is sometimes called the "uncanny" feeling. It is the fear we feel when something almost seems normal, but something is off. Horror stories use this by showing objects or situations that twist our expectations.

Some horror movies scare us by showing how fragile life can be. They might show ordinary people going about their day, then something terrible happens. The sudden change from normal life to danger can feel shocking. We fear this because we realize it could happen to anyone. Even in places that feel safe, something can go wrong. This taps into the basic human fear of losing control or facing danger when we least expect it.

Fear is also tied to imagination. Children may fear monsters under the bed. Adults might fear losing their jobs or facing a disaster. Our imagination can create worst-case scenarios that seem bigger than reality. Horror movies use imagination by showing us things we never see in daily life. These could be mythical creatures, cursed houses, or strange curses. When we watch these stories, our imagination runs wild, and we feel real fear over unreal things.

In some cases, people watch horror to understand their own fears. They face them through the screen. It is like they are testing themselves to see how much they can handle. When the movie is done, they might feel a sense of relief or even pride that they made it through. However, not everyone enjoys this feeling.

Some people have strong reactions that make them avoid horror. It depends on personal preference and how each person deals with stress.

Fear can also bring people together. When a group watches a scary movie, they share the same tense moments. They might cling to each other or gasp together. Afterward, they chat about the scariest parts. This shared experience can help friends or family bond. Fear, in a controlled setting, can make people feel closer. It is similar to how friends might go to a haunted house attraction and scream together, then laugh afterward.

There are different levels of fear, from mild unease to complete terror. Horror movies vary in how strongly they aim to frighten us. Some might just give a spooky mood, while others might be very intense with jump scares and disturbing images. People pick the type of horror they can handle. If a film is too intense, it might cause stress or nightmares. If it is too mild, it might feel boring to some viewers. Finding the right level of fear is part of why the horror genre has many subtypes.

Fear does not only come from monsters. It can also come from a sense that the normal world is twisted. Films that show normal people doing terrible things can be scary because it suggests evil can exist next door. It can even be a friend or a family member. This idea can be more frightening than a mythical monster, because it feels more possible. We might think, "Could that happen in real life?" This possibility can stick with us long after the movie ends.

Some fears are shared by many people. For instance, fear of the dark, fear of heights, fear of enclosed spaces, or fear of certain animals. Horror movies often use these common fears. A scene might place a character in a tight tunnel or a towering ledge. This is called tapping into universal fears. Since many people feel uneasy about these things, the film can scare a wide audience. Other fears are more personal. A viewer might see a scene that reminds them of a personal memory, and that can be very powerful for them, though others might not react the same way.

When we talk about why we fear, we should also note the difference between real danger and fictional danger. In real life, if a bear is chasing us, that is actual danger. We either run or find help. In a horror movie, the danger is on the screen. Our body might react, but we know we are not in real trouble. This gap between real and fictional threats lets us watch scary films without the same level of panic we would have if the threat were real. Our body still produces adrenaline, but our mind knows we can walk away safely.

Sometimes, fear helps us learn about ourselves. If a person finds they cannot watch a film about haunted houses because it is too unsettling, they have discovered a limit. Another person might learn they can handle intense gore but get very afraid of ghostly images. These differences remind us that fear is personal. What scares one person might not bother another. Horror filmmakers try to include a range of scary elements so they can reach a broad audience.

Humans might also fear things that look human but are not quite right. An example is a puppet or mannequin that appears to watch us. Because they look almost real, our mind tries to treat them like living beings. But we also know they should not move. If they do, we feel shock and dread. This is another example of the "uncanny" feeling. Horror movies that feature living dolls or eerie masks tap into this fear. It is not just about the threat; it is about how unnatural it feels.

Our beliefs can affect our fears. A person who believes in ghosts might be very scared by a movie about hauntings. A person who does not believe in them might find the same movie less scary. Similarly, a person who thinks evil spirits can possess people might be afraid of films about possession. These beliefs come from culture, religion, or upbringing. Horror movies often draw from myths and legends. If you grow up hearing that a certain kind of creature is real, you might be much more frightened if it appears in a film.

Fear can also reflect events in the world. When times are uncertain or dangerous, people might feel more afraid. Horror movies made during such times might show threats that mirror real concerns. This can add a layer of reality to the fear. For example, if people worry about disease, a film about an infection that turns people into monsters might feel extra frightening. Horror sometimes acts like a window into the collective worries of a society.

Children experience fear differently than adults. They might not understand that a movie is fake. They could think the monster is real. That is why many parents do not let young children watch scary films. Over time, children learn the difference between fantasy and reality. They may still get scared, but they learn to deal with those feelings. Adults can also get frightened, but they have more life experience to remind them that a film is fiction. Still, a movie that shows something very realistic, like a home invasion, can scare even an adult because it feels like it could happen.

There is also a type of fear that comes from tension. We know something bad might happen, but we do not know when. The movie shows a character walking through a dark room, and the music grows louder. Our heart beats faster because we expect a sudden event. Then maybe the threat jumps out, or maybe the tension is broken in another way. This tension-based fear can be very effective. It keeps us on the edge of our seat. This style of horror does not always rely on gore or monsters. It uses suspense to create fear. That is the feeling we get when we know something is lurking around the corner, but we cannot see it yet.

Another aspect of why we fear is the social factor. Sometimes, we fear being left out or ridiculed if we admit we are scared. In a group of friends, if everyone wants to watch a scary movie, one might agree to watch it even though they are very afraid. They do not want to seem weak. This can make the fear even stronger, because now they feel anxious and pressured at the same time. Horror movies can tap into the fear of not fitting in, though that is more about peer pressure than the movie itself.

Our environment can also affect how we respond to fear. Watching a horror movie in a bright living room with friends might feel less scary than watching it alone in a dark basement. The setting can make us more or less open to feeling afraid. That is why horror events often take place at night, and theaters dim the lights. They want us to be in a space where we focus on the movie without distractions, making it easier to feel the tension.

Fear also appears in our dreams. After watching something frightening, we might have nightmares. Our mind might replay the scary scenes or twist them into new forms. This is a sign that the film left a strong impression. Nightmares can feel real and can linger even after we wake up. Some people do not mind because they forget the dream quickly. Others might avoid horror because they do not like the idea of losing sleep over nightmares.

Overall, fear is complex. It can protect us, excite us, or hold us back. Horror movies use that complexity to give people a strong emotional experience. They show us strange creatures, tense situations, and dark mysteries. Our minds and bodies react by feeling worried, jumpy, or tense. Then, when the movie is over, we return to normal life. For a while, we might keep the lights on or check the closet. But eventually, we remember it was just a film. Even so, the memory of the fear might stay with us, reminding us how powerful this emotion can be.

In conclusion, we fear because our bodies and minds are wired to respond to danger, whether real or imagined. Horror movies tap into this response by showing us terrifying images and sounds. The unknown, past experiences, learned warnings, and the uncanny all play roles in how we process what we see on the screen. Fear can be thrilling in small doses, but too much can be overwhelming. Each person has their own threshold for scary content. By understanding why we fear, we can appreciate how horror movies can affect us, and we can make choices about what to watch based on our comfort level.

CHAPTER 3: HOW THE BRAIN REACTS

Our brain is a complex organ. It handles our thoughts, emotions, and actions. When we watch a horror movie, certain parts of the brain activate to deal with what we see and hear. These parts include the amygdala, the prefrontal cortex, and the hippocampus. Each plays a different role in how we process fear.

The Amygdala and Fear

The amygdala is often linked to fear. It is a small, almond-shaped region deep in the brain. When we sense a threat, real or imagined, the amygdala is one of the first areas to respond. It decides if something is dangerous. In a horror movie, a scary image or sudden loud sound might trigger the amygdala. We might flinch or feel our heart beat faster. This reaction happens quickly, even before we think things through. The amygdala can make us feel startled almost instantly.

For example, imagine you are watching a scene in which a creature leaps out from the shadows. Before you have time to realize it is just a film, your amygdala sends signals to your body to tense up. You might gasp or jerk in your seat. This automatic response is part of our survival system. In real life, it helps us react to danger, such as jumping away from a falling object. In a movie theater, it just gives us a scare, but our body reacts the same way at first.

The Prefrontal Cortex: Thinking and Control

While the amygdala sets off the alarm, the prefrontal cortex helps us think logically. It is located at the front of the brain. This area allows us to decide if we should stay afraid or calm down. Once we realize we are just watching a movie, the prefrontal cortex can send signals that say, "It is not a real threat." This helps slow our heart rate and ease tension.

However, horror movies are designed to keep us on edge. Even though part of our brain knows it is fiction, the amygdala can still stay active whenever scary scenes appear. The prefrontal cortex might remind us, "This is just acting," but the sights and sounds can still feel vivid. This tug-of-war between the emotional part (the amygdala) and the thinking part (the prefrontal cortex) is why horror movies can excite and scare us at the same time.

The Hippocampus: Storing Memories

The hippocampus is the part of the brain linked to forming memories. When something frightening happens in a movie, the hippocampus helps store that event in our mind. This is why we might remember a scary scene for a long time. Our brain considers fear important. If we are scared, it must be worth noting, in case it helps us avoid danger later.

Sometimes, the memory of a scary movie can linger for days. You might see a dark hallway at home and think about a scene you watched. That is the hippocampus at work. It has stored the details of the frightening moment, and it pulls them up when your surroundings match that memory. This is also why horror fans can recall many scary scenes and talk about them with friends. Those moments made a big impression on their brains.

Fight-or-Flight Response

When the amygdala senses danger, it can trigger what is known as the fight-or-flight response. This reaction prepares us to either confront the threat or run away from it. Our body releases chemicals like adrenaline. This can cause our heart to pound, our muscles to tense, and our breathing to quicken. Blood flow to certain areas may increase, helping us move fast if needed.

Of course, in a movie theater, we are seated and safe. We are not going to fight the monster or run away from it. But the body's reaction can still happen on a smaller scale. We might grip our seat or jump. After the scene ends, we might laugh at ourselves for reacting so strongly. That release can feel good because once the threat is over, the body starts to relax.

Why Some People Enjoy the Rush

Some viewers get a thrill from this surge of adrenaline. They might feel a rush of energy when they are startled. This rush can be exciting as long as we know we are not in real danger. It is similar to riding a fast roller coaster. Our body and mind go through quick ups and downs. Once the ride ends or the scary scene is over, many people feel relief. This contrast between tension and calm can be fun for them.

Others do not enjoy this rush. They find it too stressful. Their brain stays on alert even after the movie ends. They might feel uneasy for a long time, thinking about what they saw. Their amygdala remains somewhat active, and the prefrontal cortex cannot fully quiet those fears. That is why horror is not pleasant for every person. We each have different levels of comfort when facing scary material.

Role of Neurotransmitters

Neurotransmitters are chemicals that send messages in the brain. Two important ones linked to fear and pleasure are dopamine and serotonin. When we watch a scary film, our brain may release more dopamine. This can be tied to excitement or anticipation. It is sometimes called the "reward" chemical, though it also appears when we feel stressed. For some, this mix of stress and excitement becomes linked to enjoyment. They may remember the thrill and want to feel it again, which is why they seek out more horror movies.

Serotonin can help regulate mood and anxiety levels. If our brain is low on serotonin, we might feel more anxious. A very tense horror movie could be overwhelming for someone in that state. Our individual brain chemistry affects how we react to fright. That is why two people can watch the same film and have very different responses. One might find it fun; the other might feel too disturbed.

Desensitization Over Time

When people repeatedly watch horror, they might become less sensitive to certain scares. This is called desensitization. For instance, the first time someone sees a ghostly figure on screen, they might be terrified. But if they watch a lot of ghost movies, they might not feel the same shock after a while. Their brain has adjusted, and the amygdala no longer reacts as strongly.

Desensitization does not mean people stop being scared altogether. It varies from person to person and can depend on the type of horror. Someone could get used to jump scares but still be startled by very graphic images. Another person might not mind blood but could be unsettled by eerie sounds. The brain can adapt, but not always in the same way for every type of horror element.

Stress and Relaxation Cycles

Horror films often follow a pattern of building tension, then giving quick moments of relief. This creates a cycle: our heart rate goes up, then goes down, then up again. This roller coaster pattern can keep the brain engaged. The viewer never knows when the next scare might happen. The director might show a quiet scene to make us think everything is calm, then suddenly introduce a loud noise or a frightening figure.

These ups and downs can keep the brain alert. They also help the movie's story feel more intense. If the tension were constant with no breaks, viewers might become numb. If there were no tension at all, the movie might feel dull. The contrast is what holds our attention. Our brain responds well to changes in stimuli. That is why many scary films play with pacing. They keep us guessing when the next shock will occur.

How Age Matters

Children might have stronger reactions to horror because their prefrontal cortex is still developing. They have a harder time distinguishing fiction from reality. The amygdala, however, responds strongly from a young age. If a child watches a very scary movie, they might have nightmares or believe the threat is real. Adults usually have more fully developed thinking areas in the brain, so they can calm themselves faster. Still, certain movies can frighten adults just as easily, depending on personal triggers.

As people grow older, they might gain more control over their fear responses. They learn to remind themselves, "This is just a movie," or "That monster is just an actor in makeup." Yet, even the most logical adult can still jump at a well-timed scare, because the amygdala reacts before logic kicks in.

Individual Differences in Sensitivity

Not everyone's brain reacts to fear in the same way. Some have a very sensitive amygdala and might be easily startled. Others have a more balanced response. Some people find gory scenes more disturbing, while others might be more scared by supernatural themes. This can be due to personal history, culture, and even genetics.

A person's personality can also play a role. More anxious individuals might worry about what they see on screen long after the movie is done. They might have trouble sleeping or feel jumpy in the dark. People who are more thrill-seeking might enjoy the spikes in adrenaline and look for even scarier films.

Brain Imaging Studies

Scientists sometimes use brain scans to see how people react to horror. In many experiments, they show subjects clips of scary films while measuring brain activity. The amygdala lights up during frightening scenes. The visual and auditory parts of the brain also show increased activity because the images and sounds grab attention. Researchers see that viewers who say they like horror often have more activity in brain areas linked to excitement and reward. This suggests they may process fear in a way that feels enjoyable or interesting, rather than unpleasant.

Long-Term Effects on the Brain

Watching many horror movies is usually not harmful for most adults, but it can lead to lasting memories of scary scenes. Some people might become a bit more alert at night or in certain dark places because they recall the images. In rare cases, if a person is very sensitive, repeated exposure to frightening content might raise their general anxiety level. However, most people learn to distinguish between on-screen horror and real life.

For children, parents often choose to wait until a child is old enough to handle scary content. Because a child's brain is still growing, very frightening images might cause distress that is harder to shake off. Each child is different, so there is no exact age that fits everyone. Some might handle a milder scary movie with no trouble, while others might be too rattled.

Memory Tricks and Survival

From an evolutionary point of view, remembering scary things is a survival advantage. Our ancestors needed to remember which animals or places were dangerous. If a lion roared in the distance, that memory helped them stay alive later. Horror movies trick this survival system by showing threats that are not real. Our brains still record them as dangers, at least for a short time.

This can cause us to be more cautious after a scary film. Some people might check under their bed or avoid dark rooms. We do not truly expect to find a monster, but our brain says, "Better safe than sorry." This might be annoying, but it is part of how our mind works to keep us alive in real threats.

Why Certain Images Stay in Our Minds

You may notice that certain images from horror films stay in your mind more than others. Maybe it is a creepy face or a dramatic moment. The reason is that unexpected or unusual images make a big impact on the amygdala and hippocampus. They stand out from ordinary scenes. The brain tags them as important, so they move from short-term to long-term memory more easily.

Directors know this and often include strong, memorable visuals to keep viewers talking about their films. It can be a disturbing creature design, a shocking twist, or a very tense shot. These are the scenes that fans discuss for years, because they remain stored in the brain so clearly.

Conditioning and Learned Responses

Some horror films use techniques of conditioning. For instance, a sudden scary sound might happen right after a certain piece of music plays. If this pattern happens again and again, viewers start feeling tense as soon as they hear that music, even before the scare appears. This is similar to classical conditioning, where a signal (the music) is linked to a scary event (the loud sound or frightening image). Our brain picks up on patterns quickly.

This can lead to viewers feeling nervous any time they hear similar music outside the movie. Even if the context is different, the brain might recall the link and trigger a mild fear response. This is why certain tunes can make the hair on your neck stand up, because the brain connects them to frightening memories.

Dealing with the Aftermath

After watching a horror movie, our brain might keep replaying certain parts. Some people handle this by talking it out with friends or reminding themselves it was fictional. Others might watch something lighthearted to break the mood. This helps shift the brain out of fear mode.

In time, most scary memories fade or lose their power. The hippocampus retains them, but the emotional intensity attached to them can shrink. That is a normal process of memory storage. However, if someone keeps thinking about a horror film and feels distressed, it might help to take a break from scary content. Everyone's brain is unique, so there is no single rule for how much horror is too much.

Positive Uses of Fear Reactions

Some individuals use horror movies as a safe way to practice dealing with fear. They learn to calm themselves during tense scenes, slow their breathing, and remind themselves it is just a film. This can be a form of exposure. They face scary images and see that they can handle them. This does not work for every type of fear or for everyone, but some people find it helps them manage stress.

Others are drawn to the creative side of horror. They like seeing how directors trick the brain, how makeup artists create scary creatures, or how music shapes our emotions. They study the craft behind the scares, and this understanding can lessen the fear. When you know how a scene was made, it might feel less threatening. You can think of the lights, props, and editing, instead of focusing on the frightening story.

CHAPTER 4: OLD TRADITIONS OF SCARY STORIES

Humans have been telling scary stories for a very long time. Long before movies, people shared tales of ghosts and monsters around fires or in quiet corners of their homes. These stories passed from generation to generation through spoken words. They often warned listeners about threats, explained strange events, or simply entertained people with eerie details. Many of these old tales influence the modern horror genre. The movies we watch today owe much to these older traditions.

Early Myths and Legends

Many cultures have tales that go back hundreds, even thousands, of years. Some revolve around monstrous beasts, such as giant serpents or fierce dragons. Others focus on spirits of the dead that wander the earth. These legends grew from attempts to explain things people did not understand. For example, if crops failed for no clear reason, some might blame an angry spirit. If someone vanished in the woods, the community might speak of a lurking creature there.

These early scary stories taught people how to behave. They served as warnings. Perhaps there was a tale of a figure in the forest who punished those who stole or lied. Young children, hearing this story, would learn not to steal. Fear of the unknown kept people cautious. Over time, these stories became a core part of each culture's folklore.

Oral Storytelling

In many old communities, there were storytellers or elders who knew many tales. They would share them at gatherings, often at night. Darkness helped set the mood, and the flicker of firelight made shadows dance on the walls. The atmosphere would fill the listeners with fear and excitement. Without modern technology, these tales relied on the skill of the storyteller's voice and expressions. A sudden whisper or a loud cry could make the audience's hearts pound.

Because these stories were passed by word of mouth, details could change over time. One region might add a twist about a shape-shifting spirit, while another region might focus on curses. Despite these differences, the heart of each tale remained the same: a frightening event meant to teach or shock. These oral traditions are the root of what we now see in many horror films, which often use local legends or rumors as inspiration.

Scary Festivals and Customs

Many cultures held events during certain times of the year to confront or honor the dead. In some places, people believed that on specific nights, the barrier between the living and the dead grew thin. They lit candles or wore masks to scare away harmful spirits. These customs created a backdrop for scary tales. People would gather at dusk and tell stories of ghosts visiting from beyond. Over time, these traditions shaped how communities viewed fear and the supernatural.

In some areas, stories were told to keep children safe. They might say a spirit roamed the streets after dark, so children should stay indoors. Though it might seem like a simple scare tactic, it showed the power of storytelling to shape behavior. When horror movies began to be made, they often borrowed from these customs. Scenes of haunted nights or ghostly apparitions connect to these old beliefs that the dead could cross into the world of the living.

Folk Heroes and Villains

Some scary stories include a hero who defeats a monster. Others focus on a villain who frightens the whole town. In older tales, heroes might rely on cunning or a special talisman to ward off evil. Villains could be witches, sorcerers, or dark creatures who prey on humans. These stories gave clear examples of right and wrong. The hero stood for good, the villain for evil, and the audience learned about moral choices.

In modern horror films, we still see this pattern. The main character often faces an evil force and must find a way to stop it. This echoes the old folk tales in which a brave soul confronts a beast. Over time, the specifics have changed. Instead of swords and potions, we might see modern tools. But the idea of a hero vs. a dark threat remains strong in horror stories.

Cautionary Tales

Many old scary stories were cautionary. They showed what might happen if one broke social rules or acted selfishly. For instance, a tale might warn against wandering alone at night. If a traveler did so, they might vanish, taken by a ghostly figure. Another tale might warn against speaking ill of the deceased, lest their spirit come for revenge. These cautionary tales were a way to teach respect and caution through fear.

Modern horror movies often include similar messages. They might show a group of friends who explore a forbidden place and suffer consequences. Or they might show someone who ignores warnings about a cursed object. The story punishes those who do not listen. This pattern connects back to very old storytelling methods, reminding us that certain behaviors have risks.

Shared Themes Across Cultures

Even though the details vary, you can find many shared themes in old horror stories around the globe. One common theme is the restless spirit. Many cultures tell of ghosts who cannot find peace because of unfinished business. They wander among the living to seek justice or to express sorrow. Another theme is a man-eating creature in the woods, which warns people not to venture too far. There are also tales of a person who made a deal with dark forces and must pay a terrible price.

When horror films started being produced in different countries, these themes carried over. You might see a Japanese film about a vengeful spirit, and recognize elements of older Japanese folklore. You might see a European film about witches, inspired by centuries-old beliefs. Even if the costumes and settings look different, the core ideas—fear of the unknown, fear of punishment for wrongdoing—remain consistent.

Role of Superstition

In the past, knowledge of science was less widespread. People often explained unusual events with superstition. If livestock died suddenly, some might blame a curse. If a person acted strangely, some might suspect a witch. Such beliefs inspired many scary stories. Though modern science explains many of these events, the old tales have not vanished. They live on in legends and, by extension, in movies.

Horror filmmakers might draw on these superstitions when creating a new movie. They might craft a plot around a cursed family object or a location said to be haunted. Audiences find these elements interesting because they connect to the old fear of curses or bad luck. The result is a blend of past and present, where ancient beliefs get a modern spin in film.

Written Collections of Scary Tales

When writing systems became common, some authors began recording scary stories in books. One famous example is ancient ghost stories that were written on scrolls in certain parts of the world. Later, medieval writers compiled tales about demons, witches, and spirits. These books reached more people, and readers became fascinated by the uncanny and the frightening.

Over time, popular collections such as those by the Brothers Grimm included dark tales meant as warnings. While many of those stories were not purely horror, they had gruesome or scary elements. Children who read them might feel frightened but also learn lessons about kindness, caution, and bravery. In modern times, authors like Edgar Allan Poe or H.P. Lovecraft wrote stories that shaped the horror genre in literature. Their works combined old themes with new ideas, leading to many of the motifs we see in horror films.

Influence on Early Horror Films

When motion pictures were invented, it did not take long for creators to adapt old scary stories. Early silent films showed haunted castles, mad scientists, or mythical creatures. Directors used dramatic makeup and shadows to bring these tales to life. Audiences, who already knew some of the legends, were drawn to see them on screen.

The success of these first horror films paved the way for even more. Movie studios realized that viewers liked being scared in a controlled space. They saw the thrill that old stories could bring. So they made more films with ghosts, vampires, and other creatures drawn from folklore. Even new stories had that old feeling, because they still touched on the same fears and themes that people had told around fires centuries before.

Local Legends in Modern Cinema

In different countries, local legends still play a role in modern horror. Some regions have tales of forest spirits that lead travelers astray. Others talk about haunted temples or cursed islands. Filmmakers incorporate these stories into their scripts. This gives each region's horror movies a distinct style and flavor. For instance, a Southeast Asian horror film might show a ghost based on local beliefs, while a Scandinavian horror film might depict creatures from ancient Norse myths.

These cultural elements keep horror fresh. They introduce viewers from other places to stories they might not have heard before. Fans of the genre often enjoy watching horror films from many parts of the world, to see how different legends and old stories shape the scares.

Storytelling Traditions Still Alive Today

Despite modern technology, live storytelling of scary tales still exists. People might gather during certain holidays or events to share local ghost stories. This can happen in community centers, campgrounds, or even at parties. The atmosphere is different from watching a movie because there is no screen between the listener and the teller. The story feels more personal. This tradition shows that old ways of passing down scary tales are still alive and can still frighten people.

Some communities even organize tours at night, where guides lead groups through supposedly haunted spots. Along the way, they tell legends of the area. This is another way that old traditions continue. Even if the group does not see anything supernatural, the stories alone can give them chills.

Comparing Written Tales and Film Adaptations

Many writers have adapted old scary tales into books or plays, and then filmmakers have adapted those into movies. Each step can change the story. A short folk tale might become a full-length novel or a movie script. The basic idea remains the same—a haunting, a monster, or a curse—but new characters or plot points might be added. Filmmakers also use visual effects and sounds to enhance the story, something a written tale cannot do on its own.

This is why the same legend can show up in many forms. You might read about it in a folktale collection, then see a TV special that retells it, and later find a big-budget film version. Each version might have different details, but they all trace back to that original story told long ago. Some people enjoy comparing the various adaptations to see how the legend evolves.

Moral Messages in Old Stories

A lot of old scary stories wanted to teach moral behavior. They might show a character who is rude or greedy, and this leads to a terrifying punishment. For instance, a person who disrespects a holy place might be cursed. Another who betrays a loved one might end up haunted. These morals guided communities on how to behave. People learned to respect traditions, care for family, and not stray from accepted norms.

Modern horror can still carry moral messages, though they are not always the same. Some films show that tampering with unknown forces can lead to disaster. Others show that cruelty to others can come back in terrible ways. Even if a horror film appears to be just for scares, it might have a deeper lesson tied to old storytelling morals.

Preserving Old Stories in a New Age

As time goes on, some worry that new generations might forget the old tales. Many children spend more time with electronic devices than with elders who can share traditional stories. However, horror films often serve as a bridge. They reintroduce young audiences to legends they might otherwise never hear. Directors sometimes research local folklore to make their film stand out. By doing so, they help preserve those tales, even if they reshape them for a modern crowd.

In some places, schools or cultural groups promote activities where kids listen to or write scary tales based on local myths. This helps keep the old traditions from fading. Movies, books, and live storytelling sessions work together to ensure these stories continue. The old tales remain a rich source of inspiration for horror creators.

Differences Between Past and Present

While old stories were often linked to real-life fears (dangerous animals, unexplained illnesses, or curses), modern horror can also tap into new worries (technology gone wrong, space aliens, etc.). Still, the core fear of the unknown remains. We might no longer fear the same forest spirits our ancestors did, but we can relate to the dread of encountering something we do not understand.

Modern audiences have many entertainment options. Yet the pull of a scary story remains strong. Even with all our scientific knowledge, we can still feel unsettled by a strange figure or a spooky noise. That sense of wonder and fright connects us to our ancestors who huddled around fires, telling each other tales that made them shiver.

Folktales That Became Global

Some folk stories have spread beyond their original regions. Vampires, for example, began as a specific set of myths in parts of Europe. Now, they appear in films and books worldwide. The same can be said for werewolves or zombies. As cultures mix, these legends find new audiences. Filmmakers adapt them in fresh ways, sometimes returning to older roots, sometimes reimagining them in modern settings.

This global spread of old legends is a sign that fear and curiosity cross cultural boundaries. We might not share the same customs or language, but we can still be scared by the thought of a blood-drinking creature or a person transforming under the full moon. Horror unites people through shared fears that are as old as humanity itself.

The Enduring Power of Old Tales

Old scary stories hold a unique power. They remind us that humans have long been interested in what lurks in the dark. Whether these tales revolve around ghosts, monsters, curses, or other frightening elements, they teach us about the concerns of the past. They also show us how people learned to cope with mysteries and dangers they could not explain.

These stories laid the groundwork for modern horror, providing plots, characters, and themes that still work today. They gave us the idea of a cursed place, a haunting spirit, or a brave individual fighting a dark threat. Filmmakers add new effects, but the heart of these tales remains the same. That blend of fear and fascination, which once made old communities gasp around the fire, continues to captivate us on the screen.

Conclusion

Old traditions of scary stories are a treasure of human culture. They began with spoken tales told by candlelight or firelight, carrying lessons and warnings. Over centuries, they took many forms: folklore, written collections, and eventually films. Each culture has its own creatures, ghosts, and curses, but the themes often overlap—fear of the unknown, fear of punishment, and fear of crossing forbidden lines.

Modern horror might use flashy special effects, but it still draws on these ancient roots. The same shadows that made our ancestors jump still send shivers down our spines. The same moral lessons, disguised in frightening shapes, still guide us about right and wrong. By looking at old stories, we see how deeply fear is woven into human life. We also see how creative people have always been in spinning tales to intrigue and alarm. Our modern horror films are the newest chapter in a long tradition of scaring ourselves, and it all began with those whispered tales in the dark.

CHAPTER 5: MONSTERS AND MYTHS

Monsters in horror stories can be big, small, human-looking, or entirely strange. They can come from old myths or be made up by modern writers. But what makes a creature a "monster"? Often, it is something that frightens us because it goes beyond our idea of what is normal. Sometimes, monsters are strange mixes of humans and animals. Other times, they are reanimated bodies or spirits that break the rules of life and death. Each monster has a background that explains why it is scary, whether that background is an ancient legend or a new invention.

Origins in Myths and Legends

Many monsters in horror films trace their roots back to old myths. Myths are stories from the past that helped people explain events they did not understand. For example, if villagers saw strange shapes in the forest, they might make up a story about a huge creature that roams the woods at night. Over time, such stories became legends people told around fires, warning each other of the dangers lurking just beyond the light.

A common example is the vampire. Legends of blood-sucking beings appear in different parts of the world under various names. In some places, these creatures are said to take on the form of a bat at night. In others, they look like pale humans with sharp teeth. The fear around vampires stems from the thought of a being that feeds on human blood and might turn its victims into more vampires. This idea taps into our fear of losing control over our bodies and being attacked by something powerful in the dark.

Likewise, tales of werewolves come from ideas about people cursed to transform into wolves when the moon is full. Long ago, people often feared wild animals. Wolves, in particular, could threaten livestock or travelers. The idea of a person turning into a wolf was both unsettling and strange. It also raised questions about what happens when our animal side takes over. Modern horror films about werewolves use makeup and digital effects to show painful transformations, highlighting the frightening idea of a body twisting into a beast.

Zombies also come from older ideas, though they vary by region. Some legends talk about the dead rising from graves, either because of black magic or a curse.

In certain stories, a sorcerer controls the body of the deceased. Modern zombie tales often swap magic for viruses or other scientific causes. Still, the core fear stays the same: a lifeless body that returns to walk among us, driven by hunger. This goes against the natural order of life and death, which is why it can disturb us so much.

Monsters tied to specific lands also shape horror movies. Sea monsters in coastal regions, giant birds in areas with large cliffs, or reptilian beings in swampy landscapes reflect each culture's environment. These creatures become symbols for local fears. For instance, if a land had many earthquakes or volcanoes, stories might arise about huge beasts sleeping underground, shaking the earth when they move. Such myths explained nature's dangers before science offered its own answers.

Symbols of Deeper Fears

Beyond simply being scary, monsters often stand for bigger worries. A vampire might represent a fear of disease or fear of loss of will. A werewolf might reflect the idea that every person has a wild, uncontrollable side. Zombies can show the dread of society collapsing, with people reduced to mindless hordes. By putting a name and face to our anxieties, these monsters help us understand them in a story form.

Some monsters can also embody moral lessons. A creature might appear to punish those who lie or harm others. This echoes how old cautionary tales worked, teaching people to follow rules or face grim consequences. In these stories, the monster is not just a random terror; it is an answer to wrongdoing. Modern horror films sometimes keep this tradition. They show how a person's cruel actions bring forth a supernatural beast that targets them. Viewers might see these lessons and think about how harmful acts can have dire outcomes.

Blending Creatures from Different Cultures

As travel and media bring cultures closer, monsters from distant lands mix together in new stories. Modern storytellers might take a European vampire myth and combine it with an Asian ghost legend. This blend can create fresh creatures that feel both familiar and new. For instance, a filmmaker might imagine a vampire who not only drinks blood but also has traits of local spirits known for shape-shifting. This mixing can add fresh twists to old ideas and spark curiosity among viewers.

However, some classic monsters remain mostly the same no matter where the story is set. A mummy, tied to ancient tombs, usually appears with bandaged wrappings in Egyptian settings. This monster taps into fears of disturbing the dead and breaking curses placed on tombs. While modern films might change how the mummy moves or attacks, it still represents the danger of tampering with sacred places. Even if a new mummy story is set in another country, it often keeps the theme of ancient curses and hidden secrets.

Monsters as Reflections of Real Issues

Sometimes, a monster stands for a real issue in the world. A giant creature in a horror story might represent pollution or nuclear fear. For example, in some well-known giant monster tales, the beast is born from radioactive waste. This draws a link between human mistakes (like toxic spills) and nature's wrath. The monster becomes a symbol of guilt over harming the environment.

Similarly, a story about a mad scientist who creates a terrible beast might reflect worries about science going too far. People might wonder if we will lose control of our own creations. Frankenstein's monster is a classic example: it is built from the parts of the dead and given life through experiments. The tale warns us about the cost of playing with forces beyond our understanding. Even though it is a fictional creature, it makes us think about ethical limits in real research.

Shape-Shifters and Hybrids

Shape-shifting monsters are especially intriguing because they look normal at times, then switch form to become fearsome. Werewolves are the most famous example, but there are many others. Some creatures can appear as humans, then turn into large cats, snakes, or something else. This idea shows a fear of deception: a person we trust might be hiding a terrifying secret. Horror films use shape-shifters to keep viewers guessing about who is safe and who is dangerous.

Hybrids are another type of monster. They combine features of humans and animals in a way that seems unnatural. A creature might have a human face and an animal's body, or vice versa. This breaks the boundary between species, leaving viewers unsettled. Hybrids can raise questions about identity. If a monster is partly human, does it have feelings like a person, or is it purely a beast? This gray area can be haunting, because it challenges what makes someone truly human.

Modern Monsters

Along with old myths, we have modern monsters created by new fears. For instance, some horror stories feature machines that gain intelligence and turn against humans. Others present twisted versions of the internet, where a digital "creature" or virus can harm people. While these threats might not fit the classic image of a fanged beast, they serve the same purpose: they spark fear by showing us a threat that breaks the normal rules.

Even modern cryptid tales exist, focusing on creatures rumored to live in remote places, like large ape-like beings or giant lake serpents. Sometimes, these tales blend science with myth, suggesting there might be an undiscovered species out there. Horror movies about cryptids often show people venturing into hidden regions, only to find something that defies logic. These stories feed on the idea that we still do not know every corner of the world, and something bizarre might be waiting.

Monsters in Urban Legends

Urban legends are contemporary stories that pass from person to person, often in whispered tones. They might involve a strange figure seen on a lonely road, a hidden experiment in an abandoned building, or a ghost that appears in a city park at night. Many of these legends become modern horror tales. People share them online, making them grow quickly. Filmmakers sometimes notice a popular urban legend and decide to turn it into a movie.

These legends may not have roots in ancient myth, but they fill a similar role. They explain bumps in the night or strange shadows out of the corner of our eye.

They also carry warnings, telling people to be wary of certain places or behaviors. While older myths might talk about curses placed by ancient gods, urban legends might talk about technological mishaps or twisted accidents. But the result is the same: a monstrous presence that scares us.

How Monsters Evolve in Films

Directors regularly redesign monsters to keep viewers interested. A vampire in an older film might wear elegant clothes and speak politely, while a modern vampire might dress casually and hide among everyday people. Werewolves used to appear through practical effects and costumes, but now digital effects can show every bone crack and muscle shift. Even classic movie creatures like giant reptiles or aliens get updates with more realistic skin or new backstories.

In addition, films often show monsters that adapt to modern settings. A ghost might no longer haunt a dusty mansion; instead, it could haunt an apartment tower with flickering neon signs outside. These changes reflect the world viewers live in. The core monster idea remains, but the environment shifts to match current lifestyles. This gives the story a fresh feel, even if it draws on well-known myths.

Why Some Monsters Stand the Test of Time

Certain monsters never seem to go out of style. Vampires, werewolves, and zombies continue to appear in new films every year. Part of the reason is their flexibility. A vampire can be shown as a lonely soul longing for human connections, or as a savage beast. Werewolves can be sympathetic figures cursed to become violent once a month, or they can be unstoppable predators. Zombies can be slow and mindless or fast and strategic, depending on what the filmmakers want.

This variety means audiences can see many versions of the same monster, each tapping into different fears. The unchanging trait is the tension they create: these beings are not normal. They break rules about how living things function, so we have to pay attention. Their continued presence in horror shows they still scare us, even if we know they are fictional. These classic monsters reflect deep fears about death, nature, and our own hidden urges.

Making Monsters Realistic

Modern technology allows filmmakers to make monsters look more realistic. Computer-generated imagery can add details like scales, fur, or slime. Animatronics and advanced puppetry can help actors interact with monsters on set. This realism can heighten the fear factor because viewers see every movement in a detailed manner. It is more convincing than the rubber suits of old. However, sometimes an overuse of digital effects can make a monster seem less scary if it looks too much like a cartoon.

Many horror fans appreciate practical effects, like special makeup or robotics, because they have a physical presence on screen. A zombie with real makeup can give a sense that something truly rotting is nearby. A mechanical werewolf can cast real shadows and let actors respond to it in the moment. Even if these methods look less perfect at times, they can feel more natural. In any case, the goal is to bring the monster from myth into the viewer's reality for a short time.

Moral Complexity of Monsters

Some stories show monsters as purely evil, harming people without reason. Others present them with a sense of sadness or confusion. This second approach can make the monster more complex. For example, a vampire who regrets having to feed on humans might feel torn. A werewolf who hates the pain of transformation might isolate themselves during a full moon to avoid hurting others. These insights can make the monster sympathetic and lead viewers to ask, "Who is the real villain?"

In some films, humans can act more monstrous than the creature. This happens when characters show cruelty toward each other, while the beast is simply following its nature. The contrast between cruel human deeds and an instinct-driven monster can be thought-provoking. It suggests that being "human" is not always a sign of kindness. This can flip the script on how we think about good and evil in horror.

Modern Takes on Folklore

Filmmakers sometimes take a well-known myth and rewrite it completely. Instead of the familiar vampire who sleeps in a coffin, they might depict one that walks in daylight but is harmed by modern technology. Instead of a ghost tied to an old house, they might show a spirit tied to a website or a digital device. These changes aim to surprise viewers who think they know everything about classic monsters.

Such retellings can spark debate. Some fans like seeing new spins on old creatures, while others prefer the traditional forms. Either way, it shows the ongoing life of myths and monsters. They do not stay frozen in time. As society changes, the monsters we create or adapt also change, reflecting new inventions or fears.

Humor and Monsters

Not all monsters are used strictly for horror. Some films mix comedy and fear. A bumbling zombie or a silly vampire might appear in a spoof that pokes fun at scary tropes. This can help audiences relax around monsters they usually find terrifying. Yet even in funny takes, the monster can still have a creepy side. This blend of humor and horror might appeal to people who want a scare but do not want to feel too unsettled.

Other times, comedic moments in a horror film highlight how absurd a situation is. A group might joke to cope with their fear when facing a giant creature. This can make the monster scarier when it attacks again, because the laughter turns to screams. Mixing humor and fear can be tricky, but it can also make the film memorable.

Iconic Monster Designs

Some monster designs become so famous that fans recognize them immediately. A hunched figure with bolts in its neck might remind viewers of Frankenstein's monster. A tall, pale vampire with a high collar might bring to mind a classic count. These visual cues come from early movies and have stuck in popular culture. Even if a later film changes the design, audiences often remember the original look that set the standard.

Masks are another key part of monster design. A blank, expressionless mask on a human figure can be deeply unsettling. It hides any sign of emotion, turning a person into a mysterious form. Slasher films often use simple masks to make the villain more threatening. Likewise, animal-like masks can give the impression of a half-human beast. These props contribute to a monster's identity and help viewers feel the fear more strongly.

Why We Are Drawn to Monsters

Despite their frightening nature, monsters fascinate many people. We are curious about what breaks the rules of nature. A monster can stand for things we do not fully understand about the world or ourselves. It can also be exciting to face a horrifying concept, all while knowing we are safe in our seat. Some people even root for the monster, finding it more interesting than the human characters. Others focus on the question of how the heroes will stop it.

Collectors might buy figures or artwork of famous monsters. They may dress as these creatures during costume events. This might look odd, but it shows how a monster, once feared in a story, can become a fun symbol in everyday life. People might enjoy the thrill of wearing a werewolf mask or a vampire cape, acting out the monster's role in a harmless way.

New Myth-Making

With the spread of online storytelling, fresh monsters appear all the time. People share short horror tales about mysterious beings in chat rooms or on social media. When these new creatures gain popularity, they can become the basis for movies or shows. This modern style of myth-making is quick compared to older times. Instead of hundreds of years, a story can go viral in days. If enough people find it scary or compelling, it becomes part of our shared monster library.

This shows that we never stop inventing creatures to frighten us. Whether it is a slender figure spotted in dark woods or a digital entity that haunts your devices, we keep pushing the boundaries of what a monster can be. Some of these new myths last, while others fade. But they all add to the wide range of horrors we can imagine.

Collective Fears and Monsters

Monsters often feed on fears that many people share. If society is worried about a certain problem—like a new disease—there might be a spike in stories about infected beings. If people are anxious about future technology, we might see monsters tied to rogue robots or artificial intelligence. Horror, in this sense, is a mirror that shows our hidden dreads. The monster becomes the face of whatever big worry is lurking in our thoughts.

This connection between society's fears and the popularity of certain monsters means we can track cultural shifts by looking at horror trends. At times, giant creatures might be popular if there is widespread concern about environmental harm. At other times, vampires might make a comeback if people feel uncertain about trust or intimacy. Monsters shift shape depending on what scares us most at the time.

Conclusion

Monsters and myths are at the heart of horror. From vampires and werewolves to zombies and giant creatures, they come from ancient stories and modern imaginations alike. Some reflect local fears or moral lessons, while others symbolize global concerns like scientific ethics or social collapse. Through movies and other media, these monsters evolve, adopting new looks, abilities, and backstories to keep scaring new generations.

Yet no matter how much they change, monsters remain powerful because they break the rules of nature in ways that make us shiver. They let us explore deep fears about death, transformation, and the unknown. Myths help us give shape to these fears, turning them into tales we can share. By watching monsters on screen, we step briefly into a place where normal rules do not apply, facing nightmares we might avoid in real life.

Though we may think of them as pure fantasy, monsters can teach us about ourselves. They can show us what we fear, how we deal with it, and how our worries shift over time. By understanding the origins and roles of these creatures, we gain insight into why horror stories grip us so strongly. They speak to an age-old part of our minds that imagines threats in the dark. And they remind us that, try as we might, we cannot fully turn away from the allure of the monstrous and the mysterious.

CHAPTER 6: HORROR GROWS IN FILMS

Horror films did not appear out of nowhere. They developed step by step as filmmakers tried new ideas and used new technology. Today, horror is a well-known genre with fans around the world. But it started in simpler ways, back when movies were still silent, and special effects were basic. Over time, directors and studios realized that scaring an audience could be both exciting and profitable. They found methods to keep viewers on the edge of their seats, bringing monsters, ghosts, and killers to the big screen in ways no one had seen before.

Early Silent Horror

In the late 1800s, the first films were short clips with no sound. These early days saw a few attempts at spooky themes, but they were mostly playful. One of the first known examples is a silent clip of a haunted inn or a dancing skeleton. The acting and sets were simple, and the camera work was experimental. Still, it hinted at the possibilities of scaring people with moving pictures.

As silent movies became longer, some filmmakers turned to scary stories from books. One famous example is "The Cabinet of Dr. Caligari," made in Germany in 1920. It had twisted sets and shadows painted on walls, creating a dreamlike look that unsettled viewers. Another is "Nosferatu," a 1922 vampire film that was an unofficial adaption of the Dracula tale. The bald-headed vampire with long fingers remains a haunting image even today. These early films showed how lighting, shadows, and odd camera angles could create tension without any spoken dialogue.

Audiences found these stories eerie. They were not just simple tricks; they sparked feelings of dread. Word of mouth spread, and more people went to see these movies. Silent horror paved the way for the sound era, showing that viewers enjoyed being frightened as much as they enjoyed comedies or dramas.

Arrival of Sound and Hollywood's Golden Age

When sound entered the movies in the late 1920s, everything changed. Sudden noises, eerie music, and screams made horror more intense. Hollywood studios saw an opportunity to adapt famous horror novels for the screen. In the 1930s, Universal Pictures produced a series of monster films that became classics. These included "Dracula," "Frankenstein," "The Mummy," and "The Invisible Man." Each movie featured a distinct monster and a star actor who brought it to life.

Viewers flocked to theaters, eager to see these strange beings move and speak. Dracula's accent, Frankenstein's groans, and the hiss of the Mummy gave them a new dimension that silent films could not match. The sets were often foggy castles or eerie labs, and the costumes were striking. These images stuck in people's minds. Over time, these Universal monsters became icons of the horror genre. They were scary, yet some audiences also felt sympathy for them, especially Frankenstein's lonely creature.

By the 1940s, studios were making sequels and crossover films with multiple monsters. Audiences could watch werewolves battle vampires or see the Mummy return again. Though these movies sometimes repeated ideas, they built a big following. They also helped shape horror's image as a genre filled with creatures and spooky atmospheres.

Shifts in the 1950s: Science and Atomic Fears

After World War II, the world saw the power of atomic bombs. This led to a new wave of horror films centered on science gone wrong. Giant ants, lizards, or spiders rampaged across cities after nuclear tests. People worried about radiation, and filmmakers used this fear to create monsters. For example, a normal creature might become huge and aggressive because of atomic energy. The idea was both thrilling and a reflection of real concerns.

These 1950s horror films often mixed science fiction with scares. Characters might be researchers trying to contain giant insects or aliens. The sets featured labs, strange devices, and men in protective suits. The message was clear: playing with powerful technology could unleash forces we could not control. Audiences saw these films as both campy fun and a warning. Some films also had themes about space travel. Aliens landing on Earth might be friendly or violent. The uncertainty about what lay beyond our planet fed into the general mood of curiosity and fear.

Meanwhile, outside the United States, directors in other countries told their own horror stories. Japan famously created giant monster films, where towering creatures stomped through cities. These tapped into the anxiety of nuclear weapons and natural disasters, reflecting local experiences of war and rebuilding. Over time, these giant creature features became a subgenre known for massive battles between beasts and armies.

The 1960s: Psychological Horror and Shocking Twists

In the 1960s, filmmakers began focusing on human psychology as a source of terror. Alfred Hitchcock's "Psycho" in 1960 changed the game. It showed that real people, not just monsters, could be horrifying. Viewers were stunned by the film's twists and the idea of a seemingly normal person hiding a dark, violent secret. The famous shower scene became one of the most iconic moments in movie history, proving that horror could be both subtle and shocking at once.

This decade also saw the rise of low-budget horror films that used creativity instead of fancy effects. Directors realized they could show tension with everyday locations and normal people. A simple hotel room or a quiet suburban street could become the stage for terror if the story was right. "Night of the Living Dead," released in 1968, introduced modern zombies. Though filmed with little money, it had a raw energy that terrified audiences, showing undead figures surrounding a small group of survivors. Its dark ending left viewers unsettled. This film also tackled social issues, hinting that the real monsters might be the humans inside the house who could not work together.

During this period, black-and-white cinematography still appeared in many horror productions, even though color film was available. The choice to use black-and-white could make a film feel gritty and real. It forced viewers to focus on shadows and contrast, which are great tools for building fear. By the end of the 1960s, horror was branching out. Some directors went for psychological dread, others for graphic shocks, and some for a mix of the two.

The 1970s: A New Level of Intensity

The 1970s took horror further into disturbing territory. Films were more graphic in their violence and more intense in their themes. This was partly because of changes in movie rating systems, which allowed filmmakers to show stronger content without having to please strict censors. Audiences were ready for bigger shocks, and the decade delivered.

One major trend was movies about possession and evil forces, such as "The Exorcist" (1973). It showed a young girl possessed by a demon, with scenes that made viewers faint in theaters. The mix of religious themes, special effects, and realistic acting created a sense that something truly evil was on screen. Other filmmakers looked at real-world crimes or claimed "true story" elements to scare viewers. "The Texas Chain Saw Massacre" (1974) showed a group tormented by a family of killers, filmed with a rough style that felt almost like a documentary.

Meanwhile, directors abroad produced strong horror as well. Italian filmmakers experimented with bright, unsettling colors and intense soundtracks, creating a style some called "giallo," which was part murder mystery, part horror. The 1970s was a time when horror directors felt they could push boundaries. They revealed darker corners of human nature, leaving behind the simpler monster stories of earlier decades. Audiences found themselves asking if the threats on screen could exist in the real world.

Slasher Films of the Late 1970s and 1980s

One of the biggest subgenres to emerge from this era was the slasher film. It usually follows a masked or mysterious killer who stalks a group of people. "Halloween" (1978) is often credited with setting many of the rules for slashers. It had a menacing villain, Michael Myers, who seemed unstoppable. He appeared in quiet suburban streets, making viewers question the safety of ordinary neighborhoods.

Soon, other slasher films followed, like "Friday the 13th" and "A Nightmare on Elm Street." Each had its own killer with a distinct look and method. These movies appealed to teens and young adults, partly because the main characters were often teenagers. Viewers saw people like themselves in danger. The films mixed simple jump scares with rising tension, and many had sequels that stretched into long-running franchises.

Though some critics found these movies too violent, they drew large crowds. They also raised debates about what was acceptable to show. Filmmakers balanced the gore with tense build-ups, using music cues and camera angles to keep viewers guessing. Despite negative reviews from some quarters, the slasher wave proved that horror could be a major box office success.

The Rise of Special Effects and Creature Features

During the 1980s, special effects advanced. Studios could make more realistic monsters, thanks to animatronics, makeup, and new technology. Films like "An American Werewolf in London" (1981) stunned audiences with a detailed transformation scene. Viewers watched a normal man turn into a werewolf in agonizing detail. This level of realism had rarely been seen before.

Other creature features used animatronics or puppetry to breathe life into fantastic beasts. Horror could now show moving, snarling creatures without relying on obvious costumes. Even everyday animals could become terrifying if shown in a believable way. The 1975 film "Jaws," though released earlier, was a big influence. It showed how a mechanical shark could frighten audiences out of the water. Building on that, 1980s horror explored dinosaurs, giant bugs, and man-eating plants, often with a mix of practical models and camera tricks.

At the same time, comedy elements sometimes slipped in. Movies like "Gremlins" (1984) mixed funny creatures with scary situations, showing a more playful side of horror. These blends broadened the genre, bringing in people who liked a lighter tone while still craving scary moments. The 1980s ended with a wide variety of horror styles, from the bloody slashers to the cartoonish monster romps.

The 1990s: Meta and Psychological Themes

In the 1990s, horror saw a shift. A film called "Scream" (1996) turned the genre on its head by making the characters aware of horror movie clichés. They even listed "rules" for surviving a slasher scenario. This clever approach appealed to viewers who had grown up on earlier horror. It made them feel in on the joke while still delivering tension and scares. The "meta" style, where a film comments on itself or the genre, became more common.

Another trend was the return to psychological horror. Films like "The Sixth Sense" (1999) focused on mood and subtle unease rather than gore. The twist ending grabbed viewers, proving that a well-crafted story could shock people just as much as graphic violence. Other 1990s horror took place in normal locations, like suburban homes, but hinted at hidden spirits or evils. Technology began playing a bigger role too, with computers and cameras featuring in plots that involved haunted videos or strange online happenings.

At the same time, some older franchises continued, making more sequels. The slasher icons of the 1970s and 1980s appeared in new installments, sometimes with diminishing returns. Audiences had become more selective, and some found the repeated storylines stale. Yet the existence of these sequels showed that horror fans were still loyal to the genre. They might complain about a film's quality, but they would still show up for the chance to be scared.

Global Horror Hits the Mainstream

Throughout the 1990s and into the 2000s, horror from outside Hollywood gained attention worldwide. Films from Asia, especially Japan and South Korea, found big audiences with titles like "Ringu" (1998) and "A Tale of Two Sisters" (2003). These stories used a slower pace and eerie atmosphere. They showed ghosts in new ways, such as a vengeful spirit crawling out of a TV screen. Hollywood noticed their success and made remakes that introduced Western viewers to these styles.

Other regions had their own takes. Spanish-language horror brought rich gothic atmospheres and emotional storytelling. French filmmakers produced intense psychological or slasher-like movies that pushed boundaries. This global exchange of horror ideas helped the genre evolve. Audiences could now access scary films from many cultures, each with unique myths and filming styles.

Found Footage and New Recording Styles

With cheaper digital cameras, some directors experimented with the "found footage" format. In these films, we watch the story unfold through a character's camera or phone. "The Blair Witch Project" (1999) made this style popular. By pretending the footage was real, the movie sparked debates about what truly happened in the woods. Viewers felt like they were part of the events, with shaky shots and incomplete angles adding to the tension.

Found footage became a staple in the 2000s. Films like "Paranormal Activity" (2007) relied on static camera setups inside a home, capturing small movements and sounds at night. The tiny budgets for these films meant huge profits if they hit big in theaters. While some viewers disliked the shaky camera work, others found it more immersive. It let them imagine these events might really happen.

The Modern Era: Diverse Stories and Elevated Horror

In recent years, horror has grown to include many styles. Some call a certain type "elevated horror," which mixes deep themes with scary elements. These movies might explore grief, family problems, or social issues in a horror framework. They use dread and symbolism rather than constant jump scares. Examples include films that tackle race relations, mental health, or the bonds between family members. While these stories can be unsettling, they also make viewers think about real-life problems.

At the same time, there are still straightforward slashers, monster movies, and supernatural tales. Nostalgia for 1980s horror leads some directors to shoot with older film techniques or set their stories in that era. Streaming services also produce original horror content, allowing for a wide range of budgets and ideas. In this environment, horror fans can pick from many subgenres: slow-burn ghost stories, violent slashers, twisted psychological thrillers, or anything in between.

Horror Franchises and Shared Universes

Studios often build franchises around successful horror hits. They produce sequels, prequels, and spin-offs, letting fans stay in that world longer. If one monster or villain becomes famous, it might appear in multiple films, plus TV shows, merchandise, or even video games. This can keep a franchise going for years, although critics sometimes complain that the quality drops as sequels multiply.

Shared universes have also come into play. A studio might link several horror films with recurring characters or a common evil force. Viewers then watch to see how each story connects. This approach mirrors what comic book movies have done, but in a horror context. While it can be exciting for fans, it also challenges filmmakers to keep the stories consistent and fresh.

Growth in Special Effects and Technology

Modern technology lets horror directors show worlds and creatures that used to be impossible. Computer graphics can render complex ghosts or huge monsters. Motion-capture suits let actors perform as creatures that look nothing like humans, with every move and facial expression captured in detail. Digital de-aging can make an actor look younger or transform them into something uncanny.

However, many argue that practical effects still have their place. Some of the scariest moments come from physical props or makeup that the actors can touch. Directors may choose a blend of real effects and digital additions to achieve the right balance. It depends on the story's style and the budget available. Big studio films can spend millions on cutting-edge effects, while independent horror might rely on creativity to make chills with less money.

Horror's Ongoing Appeal

Why does horror keep growing? One reason is that fear is a universal emotion. People in many cultures feel tension and relief when a film scares them, then ends. Another reason is that each era has different worries—wars, diseases, personal struggles. Horror can show these concerns in extreme ways. Audiences can process them through the lens of scary stories, facing them in a controlled environment.

Horror also thrives on word of mouth. A movie that shocks or unsettles people will get talked about, shared online, or recommended among friends. This buzz can help smaller films reach bigger audiences, which in turn encourages new ideas. Fans look for the next big scare, while filmmakers try to create something more intense or innovative than before.

CHAPTER 7: GREAT DIRECTORS

Horror films would not be the same without the directors who guide them. These filmmakers have strong ideas about how to scare an audience. They know how to use camera angles, lighting, music, and actors' performances to create fear or tension. Some focus on quiet, slow-building dread. Others love sudden surprises and dramatic visuals. In this chapter, we will look at several directors whose work changed the horror genre. Each brought a special style, and many inspired the next generation of storytellers.

Early Pioneers

F. W. Murnau

One of the first important horror directors was F. W. Murnau, known for the silent film *Nosferatu* (1922). He took the tale of Dracula and transformed it into a haunting story. The vampire in his film was bald, with long fingernails and rat-like teeth, very different from later versions. Murnau used shadows to make scenes scary. Because there was no sound, images had to speak for themselves. *Nosferatu* remains famous for its eerie mood and creative use of light and dark.

James Whale

Another early figure was James Whale. He directed *Frankenstein* (1931) and *Bride of Frankenstein* (1935). Whale's style combined gothic sets with the human side of the monster's story. The tall laboratory towers, the claps of thunder, and the lightning bolts all became classic horror images. Whale's work made people think about who is truly monstrous: the creature, or those who fear and reject it. Even decades later, these films are praised for mixing terror with sympathy.

Tod Browning

Tod Browning gave us *Dracula* (1931), starring Bela Lugosi as the count. With his slicked-back hair and thick accent, Lugosi's Dracula became an icon. Browning created scenes with giant cobwebs and candlelit corridors to set a gloomy mood. The film introduced many to the idea of a suave vampire with a hypnotic gaze. Browning also directed other shocking tales, but *Dracula* remains the film people link most with his name.

The Master of Suspense: Alfred Hitchcock

Alfred Hitchcock was not a horror director in the traditional sense. He made thrillers and mysteries, yet many of his works feel like horror because they inspire deep tension and fear. The most famous example is *Psycho* (1960). It did not have a supernatural monster. Instead, it showed a normal-looking person with a dark secret. Hitchcock's use of editing, especially in the shower scene, made viewers feel every slash without seeing graphic details. This trick of suggestion, rather than showing everything, became a tool many later horror directors used.

Hitchcock also played with his audience's expectations. He might build up tension in one direction, only to swerve in another. Many directors learned from him that fear can be strongest when you do not show the threat right away. By mixing clever storytelling with technical skill, Hitchcock proved horror could be smart and stylish. He also showed that an isolated setting, like a lonely motel, can become the perfect place for terrible events.

George A. Romero and the Modern Zombie

Before *Night of the Living Dead* (1968), zombies in film were often linked to mind control or magic. George A. Romero changed that. He presented reanimated corpses that wandered around, hungry for human flesh. His zombies were slow and relentless. They crowded around houses, peered in windows, and pushed against doors. This created a sense of siege. The living were trapped inside, unsure how to handle the growing horde.

Romero also used the zombie threat to comment on society. His movies hinted that people could turn on each other during a crisis. Characters often failed to unite, leading to tragic endings. This deeper meaning made Romero's work stand out from simple monster flicks. In later films like *Dawn of the Dead* (1978), he used a shopping mall setting to point to consumer habits. He was not only scaring viewers—he was making them think.

Tobe Hooper and the Power of Raw Shock

Tobe Hooper gave the horror world *The Texas Chain Saw Massacre* (1974). Although it had a very low budget, it shocked audiences. Many expected a simple slasher, but what they got felt almost like a real nightmare. Hooper used rough cinematography and unsettling sounds, including constant noises from a generator and the buzzing chain saw. The film hinted at violence sometimes without showing every detail, making people's imaginations fill in the gaps. This style felt different from the polished horror of earlier decades. It was gritty, taking place in rural America, with a family of killers who seemed far removed from normal society. The film also dared to portray a final scene of wild chaos. While not everyone could handle its intensity, it left a huge mark on the genre. Hooper later directed *Poltergeist* (1982), a haunted house film that balanced family drama with supernatural chills. Yet it is *The Texas Chain Saw Massacre* that remains his most famous work.

John Carpenter and Building Tension

John Carpenter made *Halloween* (1978), one of the earliest slasher films to gain wide success. It introduced Michael Myers, a masked figure who returns to his hometown. Carpenter's style relied on slow pacing and tense music. He composed the simple piano score himself, which has since become one of the most recognizable horror themes. The film showed that a quiet suburban street could be as scary as any dark castle.

Carpenter often used steady camera shots that followed the killer's point of view. This made viewers feel they were looking through the eyes of someone dangerous. By limiting the use of gore, Carpenter kept the focus on dread and anticipation. He also directed other horror hits like *The Thing* (1982), which featured a shape-shifting alien that mimicked its victims. In that film, Carpenter mixed horrifying creature effects with the fear of not knowing who might be infected.

Wes Craven and Surprising Twists

Wes Craven had a talent for taking horror in new directions. Early on, he made *The Last House on the Left* (1972) and *The Hills Have Eyes* (1977), both known for their shocking plots. But he became most famous with *A Nightmare on Elm Street* (1984), where a creature named Freddy Krueger invaded people's dreams. By blurring the line between dream and real life, Craven introduced a fresh scare:

nowhere is safe if sleep itself becomes dangerous.

Later, Craven directed *Scream* (1996), which teased horror clichés and had characters who knew the "rules" of slasher films. This self-aware style felt clever and fun, but still frightening. Craven showed that horror could comment on itself while delivering real tension. He often wrote or co-wrote his scripts, so his personal voice shaped the feel of his movies.

Dario Argento and Italian Style

While Hollywood directors often get the spotlight, other parts of the world also produced standout horror creators. Dario Argento from Italy is a key example. He made films called "giallo," which combined mystery with strong horror elements. His movie *Suspiria* (1977) is full of bright colors, unusual camera angles, and a powerful soundtrack. It tells the story of a dance school with hidden supernatural secrets.

Argento's style includes intense close-ups and dramatic lighting, creating a dreamlike feel. He also uses blood in a bright, artistic way rather than aiming for realism. This approach, though shocking, seems more like abstract art than pure gore. Many modern directors praise Argento for his use of color and music to unsettle the audience, proving that horror can be visually striking as well as scary.

Sam Raimi and Creative Energy

Sam Raimi burst onto the horror scene with *The Evil Dead* (1981). It was a low-budget film about young people in a remote cabin, attacked by demonic forces. Raimi used wild camera moves, as if the evil itself was rushing through the woods. He mixed scary makeup effects with moments of quirky humor.

In sequels like *Evil Dead II* (1987) and *Army of Darkness* (1992), Raimi pushed the comedy further. Yet he never let go of the frightening side, creating a blend of chills and laughs. Raimi's style is marked by kinetic shots and odd angles. When someone is attacked or possessed, the camera might tilt or shake to reflect their confusion. This approach inspired many later filmmakers, showing that horror can be both frightening and playful.

Guillermo del Toro and Gothic Beauty

Guillermo del Toro is a Mexican director who loves blending fairy-tale elements with horror. His films often feature detailed sets, strange creatures, and stories about outsiders. *The Devil's Backbone* (2001) is set in an orphanage during a war, where a child ghost wanders the halls. It mixes real-world problems with supernatural fear. *Pan's Labyrinth* (2006) takes place in the aftermath of war, following a girl who sees magical beings that might or might not be real.
Del Toro's monsters can be beautiful, ugly, or both. They often have tragic backgrounds. He uses rich colors and designs, making each scene look like a painting. While some of his work leans more toward fantasy, it contains dark moments that scare viewers. Del Toro has a deep respect for old myths and legends, and his films feel like modern versions of fairy tales turned frightening.

James Wan and Modern Franchises

James Wan is known for creating or co-creating big horror franchises. He directed *Saw* (2004), a film about deadly traps set by a twisted figure who wants victims to value their lives. Though it launched a run of very graphic sequels, Wan's original film relied on tension more than gore.
Wan then moved on to *Insidious* (2010) and *The Conjuring* (2013), both of which deal with ghosts and demonic entities. He showed skill at building suspense in ordinary homes, using small sounds, flickering lights, or hidden shapes in the background. Many people praise his ability to craft jump scares that feel earned. He also sets up stories so that the audience becomes invested in the characters. By the time the main threat appears, viewers genuinely worry for their safety.

Jordan Peele and Social Commentary

Jordan Peele, who once worked mostly in comedy, surprised everyone with *Get Out* (2017). This film explored racial tension by telling a horror story about a young man visiting his girlfriend's family, who seem polite but harbor a shocking secret. Peele used everyday conversations and awkward gatherings to create a sense of dread. The film struck a chord with many viewers who saw how social unease could turn into a nightmare.

He followed it with *Us* (2019), about doubles that rise to terrorize a family. Peele's films often link personal fears with bigger social problems. He blends suspense, sudden violence, and thought-provoking themes. By doing so, he has introduced new audiences to horror that makes people think as well as jump.

Other Influential Names

There are many directors who left marks on horror. **Clive Barker** wrote and directed *Hellraiser* (1987), showing creatures from another dimension who mix pleasure and pain. **David Cronenberg** explored "body horror," focusing on how flesh and technology might combine in terrible ways. Films like *The Fly* (1986) show transformations that are both tragic and revolting. **Mike Flanagan** directs modern ghost stories that are often as sad as they are scary, such as *Oculus* (2013) or his series on streaming platforms. And **Ari Aster** brought unsettling tales like *Hereditary* (2018), weaving family drama with supernatural dread.

Each director adds their own voice. Some rely on atmosphere, others on shocking images, while still others use deep psychological fears. Audiences may prefer one style over another, but all help keep the genre alive and varied. By studying their works, we see how many ways there are to spark fear in a viewer.

Common Traits of Great Horror Directors

1. **Personal Vision**: These directors have a unique style or idea they stick to. Whether it is the color schemes of Argento or the dreamlike scenes of Craven, they know how to make their work stand out.
2. **Mastering Atmosphere**: Horror often depends on mood. Lighting, set design, and sound all matter. Great directors use these tools to build tension even before anything scary appears.
3. **Understanding Fear**: They know that fear can come from different sources—shock, suspense, disgust, or deeper themes. They choose which form of fear fits their story.
4. **Memorable Characters**: Whether it is a killer with a mask, a ghost with a tragic past, or a normal person in an extreme situation, great directors

make us care about what happens to them. This personal link makes the horror stronger.

5. **Innovation**: They often try new techniques. This could be fresh camera work, special effects, or plot twists. By surprising the audience, they keep horror from feeling stale.

Influence and Legacy

Many of these directors inspire young filmmakers. Film students watch classics to learn how to build suspense or frame a shot. Some directors even mentor new talent, passing on what they have learned. Others give interviews or release behind-the-scenes documentaries that reveal how they pulled off certain effects.

Viewers also become more informed over time. People watch a film like *Psycho* or *The Exorcist* and then expect new movies to bring something even more exciting. This raises the bar. Modern directors must think about what has already been done and what they can do differently. They may combine ideas from older masters with modern concerns, like social media or virtual reality.

Challenges Facing Directors

It is not always easy for horror directors to get their projects funded. Some producers may worry that a film is too strange or too unsettling. However, the success of many horror hits proves that audiences are willing to watch bold ideas. Directors who manage smaller budgets often rely on creative methods to scare viewers, such as using silence, shadows, or the power of suggestion. Directors must also handle issues of censorship in certain countries. Some places have strict rules about violence, ghosts, or other content. This can limit what is shown or force cuts to certain scenes. Great directors find ways around this, using hints or metaphors instead of explicit gore. Sometimes these limits can lead to more subtle, clever approaches.

Working with Actors and Crew

A horror director must guide actors to show fear or be frightening in believable
ways. That can be tricky if the set is full of camera equipment and lights. Actors
have to pretend something truly scary is happening. A good director creates the
right atmosphere on set, maybe keeping things quiet before a tense scene. They
might explain the character's emotion or show reference images so the actor
understands the tone.
Directors also work closely with special effects teams, makeup artists, and music
composers. They may say, "I want the monster to look sad as well as scary," or
"This scene needs a sudden burst of strings in the score." Together, they shape
the final product. Good communication is key. If everyone shares the same
vision, the film feels consistent.

Staying Relevant

Horror directors often adapt to new times and trends. The rise of found-footage
horror, for example, led some established names to try that style. Or they might
incorporate modern concerns like hackers, phone stalking, or artificial
intelligence. Directors who stay up-to-date can connect with younger audiences
while still respecting older horror traditions.
Some directors prefer to keep making the same style of films, as they have a loyal
fan base. Others experiment in different genres, then return to horror to bring
fresh ideas. The important thing is that the core remains scary. If a film no
longer frightens or unnerves people, it might not hold onto the "horror" label for
long.

Why Directors Matter

You might wonder why we focus on directors, rather than writers or producers.
Of course, many people work on a horror film. Writers create the script,
cinematographers plan the shots, editors piece the film together, and so on. But
the director is the guiding force. They decide how scenes are paced, how actors
deliver lines, how sets are designed, and how music is used. If you watch several
films by the same director, you often see patterns that show their personal style.

By learning about these directors, we see how horror movies are more than random shocks. They are crafted by people who study the emotion of fear and know how to shape it. Each director's background, culture, and taste contribute to a film's feel. Some do it in a calm, creeping way, while others go loud and intense. Either path can work if the person behind the camera has a clear vision.

Conclusion

Great horror directors are like maestros, orchestrating fear in many forms. From silent film pioneers like Murnau to modern innovators like Peele, they show that horror is always growing. Each one takes the basic idea of scaring people and spins it in a new way. Some focus on monsters, some on human evil, and some on the thin line between reality and nightmares.

Their films stay with us because they speak to hidden worries: losing control, facing the unknown, or confronting something that should not exist. As long as people have these worries, horror directors will find ways to show them on screen. By watching their works and seeing how they use imagery or sound, we gain insight into our own fears. That is the power of a skilled horror director: they turn dread into an art form, and they help us explore what it means to be afraid—and why we cannot look away.

CHAPTER 8: TYPES OF SCARY THEMES

Horror is a wide genre with many different kinds of stories. Some films scare us with monsters, others with human villains. Some prefer quiet, creeping dread; others rely on loud shocks and big reveal moments. In this chapter, we will break down common scary themes. These themes help filmmakers build fear in specific ways. Whether you like supernatural tales or realistic thrillers, there is likely a theme that appeals to you.

---

1. Supernatural Horror

Supernatural horror involves elements that go beyond normal science, such as ghosts, demons, or curses. These stories often explore what happens when the spirit world and the human world collide. The fear comes from not understanding these forces or knowing how to stop them.

Ghosts

Ghost stories usually involve a restless spirit linked to a specific place or object. They might seek revenge, warn the living, or replay a tragic past. The setting might be a haunted house, an abandoned hospital, or a lonely road. The scariest part is often seeing a glimpse of something that should not be there—a face in the mirror or a shape in the corner of the room.

Demons and Possessions

When a person is possessed, a dark force takes control of their body. This leads to unsettling behavior, strange voices, or physical changes. Possession films can be very tense because a loved one might become unrecognizable. People around them are desperate to drive out the demon, often turning to religious rituals. This raises questions about faith, evil, and the limits of human strength.

Curses and Occult

Curses are spells that bring bad luck or harm to those who break certain rules or disturb sacred places. Occult horror involves secret groups or forbidden rituals. Stories might include a book of spells, a haunted artifact, or a gathering of people trying to summon a dark force. The fear comes from stepping into mysteries best left alone.

2. Psychological Horror

Psychological horror focuses on the human mind. Instead of giant monsters or magic, it deals with paranoia, guilt, or deep emotional wounds. These stories can be slow and subtle, making viewers question what is real.

Unreliable Characters

A classic approach is to follow a character who sees or hears disturbing things. But we do not know if these events are real or a sign of a breakdown. The horror grows as we discover the character's backstory. Maybe they went through a huge trauma that changes how they see the world.

Isolation

Being alone in a remote place, cut off from help, can drive people to despair. This might happen in a snowy lodge, a cabin in the woods, or even a space station. The mind starts to play tricks. Isolation horror explores how loneliness or claustrophobia can be scarier than any monster.

Inner Demons

Sometimes, the biggest threat is the main character's own thoughts. They might carry guilt or shame, and these feelings come to life in frightening ways. The story can blur whether the horror is external or a symbol of the character's mental state.

---

3. Slasher and Serial Killers

This theme features a dangerous person—often masked—who targets individuals one by one. The fear is that a real human, not a supernatural being, is stalking victims. It is direct and can happen in settings that feel familiar, such as neighborhoods or summer camps.

Masked Villains

The mask hides the killer's face, removing any hint of emotion. This turns them into a blank shape with the sole aim of harming others. Classic examples are Michael Myers in *Halloween* or Jason Voorhees in *Friday the 13th*. The slow, steady pace of these killers can be unnerving.

Creative Methods
Slasher films often focus on the killer's methods, which might involve a
trademark weapon. Some show little blood, relying on suspense. Others are
more graphic, leaving audiences uneasy at the sight of violent acts. The key is
the chase and the question: who will escape?

Urban Legends
Some slashers link to urban legends, such as a hook-handed figure who appears
when people park in secluded spots. These stories tap into everyday warnings
and teenage fears. They also pass from friend to friend, making them seem
almost believable.

---

4. Monster Horror

Monster horror places creatures at the center of the scare. These might be
beasts from old myths or creations of science gone wrong. They can be huge,
smashing entire cities, or small and hidden.

Classic Creatures
Vampires, werewolves, and mummies are classic monsters. Their tales come
from folklore or ancient writings. Films featuring these creatures often keep the
older mood with gothic sets and dramatic flair. However, modern takes can place
them in new settings, like a city apartment or a high school.

Giant Beasts
Giant monster stories usually show a massive beast wrecking towns or battling
the military. This might reflect fears of nature rising against us, or it might link
to modern worries about toxic waste or nuclear testing.

Man-Made Monsters
Some monsters result from experiments that push science too far. The creature
might break free from a lab or spread a deadly virus. This theme warns that
humans should not meddle with forces they cannot control. Fear arises from
seeing nature strike back or science misfire in awful ways.

---

5. Body Horror

Body horror involves disturbing changes to the human form. It might show infection, mutation, or invasive procedures. The idea is to make viewers feel uneasy by twisting something we normally trust—our own bodies.

Disease and Infection
Stories where a virus spreads, turning people into creatures or zombies, fall under this theme. The focus is on the decay or transformation of flesh. Some show large-scale outbreaks, while others focus on a few people trapped with someone infected.

Mutation
Characters might slowly grow extra limbs or merge with animals or machines. Directors use special effects to show skin stretching or bones warping. This can be very unsettling, as it hits close to home. We imagine what it would be like if our own bodies betrayed us.

Parasitic Horror
This is when a creature lives inside a host, feeding on or controlling it. The film might show the slow realization that something is crawling under the skin. It combines fear of disease with the idea that we are not alone in our own bodies.

---

6. Cosmic or Lovecraftian Horror

Cosmic horror suggests that there are huge, ancient forces in the universe that are beyond human understanding. H. P. Lovecraft was a writer who made this idea popular. His stories often feature strange gods or creatures that see humans as unimportant.

Unknowable Forces
In cosmic horror, characters may find clues of an ancient evil. They realize their minds cannot grasp its true form. This leads to madness, because the knowledge is too big. Films might show only hints—a weird symbol, a giant eye in the dark—leaving the audience to wonder what else exists offscreen.

Hopelessness
A key part of cosmic horror is the sense that humanity is tiny. Even if the

characters fight back, the evil is so huge they might not succeed. This can be frightening because it flips the usual idea that heroes always triumph.

Strange Creatures
Tentacled shapes, multi-eyed beasts, and shifting masses of flesh often appear in these stories. They might come from another dimension or lie sleeping underground. The fear is both about the creature's danger and the terror of facing something our minds cannot handle.

---

7. Found Footage and First-Person Horror

Found-footage horror tries to look like real video, shot by the characters. The camera might be shaky, and scenes can cut abruptly. It makes viewers feel like they are watching raw evidence of something terrible.

Handheld Cameras
People in the story record events with a phone or camcorder. This can be a group of friends exploring a haunted place, or someone trying to document an alien sighting. The style can be immersive, as we only see what the camera points at.

Building Realism
By using amateur actors or lacking a music score, found-footage films seem more real. Viewers might wonder if the events truly happened. The technique became popular with films like *The Blair Witch Project*, where audiences were unsure if it was fiction or not.

Limited Information
A benefit of this style is that we only get as much info as the person behind the camera. If they run away in fear, the shot becomes shaky, and we might miss important details. This adds mystery and tension.

---

8. Horror Comedy

This theme mixes humor and scares, giving viewers a chance to laugh and then jump. The jokes might poke fun at horror clichés or show silly reactions to frightening events. Some people enjoy this balance because it lowers the tension before raising it again.

Playful Monsters

The creatures might act in odd ways, or the characters might bicker in comedic styles. One minute, they are cracking jokes; the next minute, they are panicking as zombies approach.

Parody

Some horror comedies parody well-known films, making fun of their serious tone while still paying tribute. Viewers who know horror tropes can spot all the references and enjoy them.

Balance of Tension and Laughter

Directors have to manage this style carefully. Too many jokes can remove the fear. Too few jokes might leave viewers confused about the tone. When done well, horror comedy can attract both fans who like to be scared and fans who like to laugh.

---

9. Tech and Digital Horror

In our modern world, technology can be scary, too. Some films focus on haunted apps, creepy websites, or AI gone wild. These stories tap into fears that we rely so much on our devices, we might be vulnerable.

Online Hauntings

Characters might encounter a strange computer file or a social media account that curses anyone who views it. As they try to warn others, the curse spreads like a virus. The horror is in how fast things can travel online.

Smart Homes

A house controlled by AI could lock its owners inside, turning the home against them. Doors, lights, and cameras all respond to the AI, leaving humans with no control. This might symbolize our dependence on technology and how it can fail us.

Virtual Reality

Some stories show people entering a VR game that becomes too real. The lines between the digital world and real life blur, causing paranoia. If you die in the game, do you die outside it? This theme links to the idea that advanced technology could trap us in illusions.

---

10. Folk and Regional Horror

Folk horror grows from local customs, legends, and beliefs. It might show ancient rituals or rural settings where old rules still hold power. The fear comes from the feeling that there is a hidden tradition the outsider does not understand.

Remote Villages

A person arrives in a distant community. The people seem friendly at first, but odd customs surface. Maybe they worship a harvest deity or hold ceremonies on special nights. The newcomer uncovers secrets that put them in danger.

Pagan Influences

Folk horror often involves pagan symbols, old gods, or nature spirits. These might require sacrifices or special offerings. The storyline suggests that ignoring these rites can anger powerful forces.

Cultural Roots

Different parts of the world have their own folk horror traditions. In one place, it could be forest spirits. In another, it might be undead warriors returning to claim their land. By using local myths, folk horror feels more authentic.

Combining Themes

Many horror films blend several themes. For example, a movie might start as a slasher, then reveal a supernatural twist. Another might seem like a simple ghost story, but it also has strong psychological layers. Mixing themes can keep the audience guessing and add depth.

Directors might also change the setting halfway through, such as shifting from a normal city to an underground world full of monstrous creatures. By doing so, they surprise viewers who thought they knew what to expect.

Choosing a Theme

Filmmakers pick themes based on what frightens them or what they think will connect with viewers. If a director grew up hearing ghost tales, they might lean toward supernatural horror. If they are worried about technology, they might make a digital horror story.

Audiences also choose which themes appeal to them. Some love the tension of slashers, others prefer the mood of a haunting. By knowing the different themes, viewers can find the type of horror that matches their taste—whether they want a quiet slow burn or a wild, fast-paced scare.

Popular Trends Over Time

Horror themes go in and out of style. At one point, slashers ruled the box office. Then, supernatural stories about haunted houses became popular. Found-footage had a big moment, and so did zombie outbreaks. Right now, many directors explore psychological drama or social issues alongside horror.

These shifts happen because real-world events change our fears. If society is worried about illness, body horror might become more common. If technology advances quickly, digital horror might pop up more often. Horror reflects what people find scary at the moment, which is why the genre never stays in one place.

Why Themes Matter

Themes give structure to a horror story. They shape the rules of the world. In a vampire film, you expect certain traits, like sensitivity to sunlight or a craving for blood. In a possession film, you expect exorcisms or talk of spirits. This framework helps viewers know what kind of fear to anticipate.

Themes also help creators stand out. A director might choose a lesser-used theme, like cosmic horror, to make their film feel special. Or they might blend two popular themes in a way no one has tried before. For example, combining folk rituals with digital technology might produce a fresh twist—ancient curses spread through modern devices.

Emotional Effects

Each theme can spark different emotions.

- **Supernatural**: Awe, wonder, and dread of the unknown.
- **Psychological**: Anxiety, confusion, and deep sympathy for characters losing their grip on reality.
- **Slasher**: Shock and adrenaline, with a strong sense of danger.
- **Monster**: Excitement at seeing a powerful creature, mixed with fear of being overpowered.
- **Body Horror**: Disgust and discomfort as characters' bodies change.
- **Cosmic**: A sense of being small in a vast universe.
- **Found Footage**: Unease from the raw, unpolished view of events.
- **Horror Comedy**: A mix of laughter and tension.
- **Tech Horror**: Worry about how our inventions might turn on us.
- **Folk Horror**: Distress at hidden customs and the threat of being an outsider.

By choosing a theme, a filmmaker decides which emotional path they want the audience to follow.

The Future of Horror Themes

As the world changes, new themes might emerge. Perhaps concerns about climate shifts will lead to stories about storms or mutated animals. Maybe advanced AI or virtual worlds will create brand-new fears. Horror has always adapted to reflect what people find scary. In that sense, the genre will keep evolving.

We might also see more merging of global myths, as cultures share stories online. A ghost tale from one region could blend with a demon legend from another, creating a hybrid theme. The possibilities are endless, as long as there are people eager to be frightened by something new or something reimagined.

CHAPTER 9: HORROR ACROSS THE WORLD

Horror is not limited to one country or region. It has grown across many places, shaped by local myths, history, and beliefs. People in different parts of the world have fears based on their surroundings—forests, deserts, mountains, busy cities, or quiet towns. As a result, each region's horror films have a distinct feel. Some might focus on spirits that come from old traditions, while others might show creatures tied to local folklore. Some places worry about ghosts, while others fear curses or witchcraft. In this chapter, we will look at how various parts of the globe produce horror that reflects their cultures.

East Asia

East Asian horror often focuses on spirits, curses, and strong emotional ties. Many of these stories come from ancient tales about restless ghosts. These ghosts might hold grudges or appear because of deep sadness. Several East Asian countries share a belief that people who die tragically can come back as spirits.

Japan

Japanese horror is famous for eerie ghost stories. Films like *Ringu* introduced the idea of a cursed videotape, while *Ju-On* (known as *The Grudge*) focused on a house marked by violent events. These stories often show pale ghosts with long, dark hair. They creep through empty rooms and appear in unexpected places. Japanese horror can be quiet and slow, using silence to scare. Sometimes, the ghost's presence grows in intensity, leading to a dreadful ending.

Japan also has a tradition of *yōkai*, or strange creatures and spirits. Some are playful; others are hostile. Modern films may adapt these old legends into new plots. People might see a figure at night that ties back to a folk story from centuries ago. Even in large cities, the idea of hidden spirits remains strong. Directors aim for an unsettling mood rather than loud jumps. Their films might reflect a sense of guilt or sorrow. The result can be haunting, focusing on the power of unresolved pain.

South Korea

South Korean horror often mixes family conflicts or social issues with ghosts or other supernatural forces. A well-known example is A *Tale of Two Sisters*, which mixes mental strain with a ghostly presence. Another is *Train to Busan*, a zombie film that also shows how people treat each other when faced with disaster.

South Korean filmmakers focus on characters' emotional states. They use this to raise tension. Viewers learn about what the characters regret or fear. When the supernatural threat appears, it ties into those inner conflicts. The fear builds not just from a scary entity, but from relationships strained by secrets. Violence can be graphic in certain works, but it is balanced by dramatic storytelling.

China and Hong Kong

Chinese horror cinema has rules about how ghosts are shown, because of local film guidelines. As a result, many older Hong Kong horror films used a blend of comedy and fear. Chinese legends include hopping vampires (jiangshi) dressed in traditional clothes, moving stiffly with arms out. Directors sometimes used comedic scenes to avoid censorship or lighten the terror.

In recent times, Chinese-language horror has grown in variety. Some stories rely on ancient curses or deals with spirits. Others take place in modern buildings. Hong Kong directors, in particular, have been known for quick pacing and sometimes over-the-top scares. Even so, they often keep a connection to local beliefs about the afterlife.

Southeast Asia

Southeast Asia includes nations such as Thailand, Indonesia, Malaysia, and the Philippines, each with its own folklore. Common beliefs involve spirits, curses, or guardian entities in nature. Horror stories might feature forests or old villages where strong beings dwell.

Thailand

Thai horror films frequently show ghosts who cling to the living because of strong emotions. A ghost might be a betrayed person seeking payback. One example is *Shutter*, about a spirit captured in photographs. Thai directors use jump scares mixed with heartfelt backstories. The ghosts are not random; they connect to a past event that caused great harm.

Another theme in Thai horror is the idea of karma. If someone does something cruel, they may face supernatural punishment. Monks or holy rituals often appear, reflecting the local religion. The tension arises from whether the main characters can escape the spirit's curse or atone for their misdeeds.

Indonesia

Indonesia has many islands with many myths. Horror films might show a *kuntilanak*, a ghostly figure of a woman who died during childbirth, or a *pocong*, a wrapped body that hops around the graveyard. These beings are well-known in local lore. Filmmakers use dark, rural settings like old houses or forests to heighten fear.

Recent Indonesian horror, such as *Pengabdi Setan* (known as *Satan's Slaves*), gained global attention. It combined family drama with eerie ghost scenes. The film had a slow build, letting viewers sense that something was wrong before revealing the full horror. Local spiritual practices often play a role, adding realism and cultural depth.

Malaysia and the Philippines

Malaysian horror includes tales of *pontianak*—a vengeful female spirit. Similar to other Southeast Asian ghosts, it often appears near banana trees or in deserted spots, luring victims with cries. In the Philippines, creatures called *aswang* roam around, sometimes shapeshifting or sucking blood. These legends shape modern films, which might set the story in remote villages or in big cities where old beliefs survive.

South Asia

South Asia, which includes India, Pakistan, and surrounding areas, has a long tradition of supernatural tales. Movies from this region often mix music, drama, and romance with horror elements.

India

Indian horror films sometimes feature ghosts who cannot rest because of a violent or unhappy past. Some older Bollywood horror productions were known for musical numbers combined with haunted mansions. In recent years, directors have tried more focused horror, with fewer songs and stronger tension.

Local beliefs in spirits and reincarnation often appear in these stories. People might fear curses left by ancestors or suffer from black magic cast by an enemy. Indian horror can also highlight social problems, such as unfair practices, but show them through scary themes. For example, a ghost might be the spirit of a wronged person seeking justice against a wealthy family.

Pakistan
Pakistani horror is smaller in number compared to Indian cinema, but there are still films and TV series that explore hauntings or spiritual threats called *jinn*. These are invisible beings mentioned in folklore. Viewers see how a household deals with unexplained events, often calling religious figures for help. The style can be slow-paced and heavy on atmosphere.

The Middle East

In many Middle Eastern countries, beliefs about spirits called jinn are common. These spirits can be good or bad, living in a parallel realm. Some horror stories revolve around what happens if a jinn attaches itself to a person.

Iran
A notable example is *Under the Shadow*, set during the war in the 1980s. A mother and child in Tehran face not only the threat of bombings but also a jinn haunting their apartment. The film uses real historical tension to enhance the fear. Western audiences found it fresh because it combined everyday life in Iran with supernatural dread.

Other Middle Eastern Regions
Horror production is not as large in some parts of the Middle East due to censorship or lower budgets. Still, you may find local folktales that inspire smaller films or online shorts. These might explore curses found in ancient ruins or spirits that roam the deserts.

Africa

Africa is vast, with many countries and cultures. Horror themes vary widely. Some stories involve witchcraft, curses, or creatures linked to tribal lore. Others focus on modern city life mixed with supernatural elements.

Nigeria

Nigeria's film industry, known as Nollywood, produces many low-budget films, some of which are horror. They might show witches casting spells or ghosts haunting families. The visual effects can be simple, but they reflect local beliefs about magic and moral lessons.

South Africa

Some South African horror films address social divides or historical events. They might use creatures or curses as metaphors for problems like inequality. Directors may set stories in rural villages or in big cities where old beliefs clash with modern life. A few also show Western-style slashers but blend them with African myths.

Egypt and North Africa

Egypt has a film history going back to the early 1900s. Though not always known for horror, it has produced movies about haunted sites or curses tied to ancient tombs. North African filmmakers sometimes weave in Islamic beliefs about jinn or curses, similar to the Middle East.

Europe

Europe has many countries, each with its own style. Gothic tales, vampire myths, and old castles are common in some places. In others, modern stories reveal deep psychological fear.

United Kingdom

British horror has a tradition of gothic style, thanks to old castles and foggy moors. Hammer Films in the mid-1900s produced many vampire and Frankenstein movies, known for bright red blood and moody sets. More recent British horror often mixes social commentary or humor. For example, *Shaun of the Dead* used zombies in a comedic way.

The United Kingdom also gave the world classics like *The Wicker Man*, a folk horror film about a strange island community. Viewers see a police officer confronted with eerie local customs. This mix of old beliefs and modern norms is a hallmark of British folk horror.

Italy

Italy has a unique horror tradition called "giallo." Directors like Mario Bava and Dario Argento used bright colors, striking music, and plots filled with mystery. *Suspiria* is a key example: a ballet academy hiding a dark, supernatural secret. These films often contain stylized violence and focus on atmosphere over realism. Italy also produces zombie or cannibal-themed movies that can be graphic, reflecting the country's flair for extreme visuals.

Spain

Spanish horror, particularly since the 1970s, has been praised for strong stories. Filmmakers like Guillermo del Toro, though from Mexico, sometimes set movies in Spain (for instance, *The Devil's Backbone*). Spanish directors also made [Rec], a found-footage film about an apartment building hit by a strange infection. Spain's horror often blends real historical or social context with scares—during times of political change, for instance, the fear in a film might echo real tensions.

Nordic Countries

Scandinavia (Sweden, Norway, Denmark, Finland, and Iceland) is sometimes known for quiet, dark landscapes. Nordic horror can be subtle, drawing on long winters, isolated towns, and old sagas. Films might show hidden creatures in the snowy forests or trolls in remote mountains. There is often a sense of loneliness or reflection on nature's power, giving a chilly, eerie vibe.

Latin America

Latin America includes countries in Central and South America, plus Mexico. Horror here can combine Catholic beliefs, local legends, and strong family ties. Curses, spirits, and vengeful ghosts are common.

Mexico

Mexico has a rich tradition of ghostly legends, such as *La Llorona*, the weeping woman who wanders at night mourning her lost children. Filmmakers continue to adapt this story in new ways. Mexican horror often merges personal drama with supernatural elements. Guillermo del Toro's early works set the stage for others, blending childhood experiences and scary fantasy.

Brazil

Brazilian horror can look at city life as well as remote jungles. Some stories focus on the Amazon region, featuring unknown creatures or curses among isolated tribes. Others might show how poverty in big cities leads to dark tales of crime or revenge. There are also Brazilian films that adapt European horror styles but add local touches.

Argentina and Other Nations

Countries like Argentina produce supernatural films tied to local myths or urban legends. For example, *Terrified* (original title *Aterrados*) explores strange events in suburban houses. It is praised for using subtle images and an uneasy mood. Many Latin American filmmakers use cramped city spaces or old neighborhoods to heighten fear, showing that the supernatural can lurk anywhere.

North America

North America, especially the United States, is home to Hollywood. Major studios produce large-scale horror that often goes worldwide. Canada and Mexico are part of this region as well, each with their own contributions, but the U.S. typically dominates in sheer volume.

United States

U.S. horror covers nearly every subgenre. From classic monster films by Universal in the 1930s to modern slashers, found footage, and paranormal stories, American horror is diverse. Big names like Stephen King's works are adapted into films and TV shows, reflecting small-town fears and supernatural events.

Cultural diversity in the U.S. also leads to stories about different myths. Directors from immigrant backgrounds might adapt legends from their homelands into an American setting. The wide range of environments—urban centers, desert highways, snowy mountain towns—offers many places for creepy events.

Canada

Canadian horror can overlap with American, but it sometimes has a quieter tone. Directors such as David Cronenberg introduced "body horror," focusing on changes in the human form. Canadian landscapes also appear in werewolf or zombie tales. The mix of English and French cultures can bring unique influences, leading to stories that feel slightly different from Hollywood standards.

Australia and New Zealand

Australia and New Zealand horror often highlights remote, rugged areas—deserts, bushlands, or far-off towns. This isolation can mean help is far away. Directors might show dangerous wildlife or mythic creatures from Aboriginal lore (in Australia) or Māori traditions (in New Zealand).

Australia
Films like *Wolf Creek* revolve around travelers encountering violent killers in the Outback. The harsh environment adds to the fear. Others might focus on ghosts in old colonial buildings. There is a sense of tension between modern life and the lingering presence of older cultures.

New Zealand
Though smaller in production, New Zealand gave the world directors like Peter Jackson, who started with gory horror comedies. *Braindead* (also called *Dead Alive*) is known for extreme splatter humor. Other New Zealand films explore rural legends or eerie coastal towns. The landscape plays a big role, offering wide, lonely views that heighten the creepiness.

Cross-Border Influences

In today's world, horror films from one country can easily reach another. Streaming services let viewers watch Japanese ghost stories, Spanish thrillers, or Indonesian demon tales. Filmmakers draw ideas from different cultures, mixing them into fresh stories. For example, a Western director might adapt an Asian ghost legend, or an Asian director might use European vampire lore in a local setting.

This exchange leads to interesting results. Some movies fuse folk horror from one region with modern city life from another. Directors might pay tribute to a classic foreign film by reusing certain visual cues or story points. The global nature of media means viewers are exposed to many forms of fear, broadening the entire genre.

Changing Tastes and Local Issues

Horror often reflects the worries people face. If a country has gone through war or social unrest, that might appear in its horror stories. For instance, some Latin American films echo the region's dictatorships or conflicts, turning them into supernatural threats. In Asia, rapid urban growth might cause anxieties about losing older traditions, so ghosts become symbols of the past.

Local censorship or cultural guidelines also shape horror's look. Some countries allow more graphic violence, while others require milder scares. Some require that ghosts be explained scientifically, so they are not shown as real supernatural beings. Filmmakers find ways around such rules, sometimes hinting at the supernatural rather than showing it directly.

Why Cultural Horror Matters

Seeing horror from many places helps us understand how people around the world think about fear. Each region's legends might teach us about old beliefs or local superstitions. A story of a cursed forest in Southeast Asia or a haunted castle in Europe reveals something about the land's history. The ghosts or monsters stand for problems that matter to the culture—like past wars, social inequality, or regrets passed down through families.

Watching these films can also break stereotypes. It shows that fear is a shared emotion. Even if the details differ, we all know what it feels like to be alone in a dark hallway or sense something is following us. Local myths might be new to a global audience, but the feeling of dread is universal. This connection helps us see that horror is not just about scaring people; it is about telling stories of what truly disturbs us.

Global Festivals and Awards

There are film festivals dedicated to horror, such as the Sitges Film Festival in Spain or the Fantasia Festival in Canada. These events spotlight movies from many countries, letting smaller productions reach a wider audience. They also prove that horror is an art form with serious fans and critical respect. Judges may give awards to directors who present fresh angles on fear, or who adapt folklore in creative ways.

These festivals can launch new talents. A chilling film made in Thailand or Colombia might get a lot of attention, leading to international deals. The growing popularity of horror streaming also helps. Platforms can subtitle or dub films, sharing them with viewers who might never have seen them otherwise.

Blending Myths: New Directions

As people move and mix cultures, we might see more horror that blends influences. A film could focus on a family with parents from different backgrounds. The child might face two sets of beliefs about ghosts or curses. This can lead to layered stories where multiple myth systems collide, creating new types of threats.

Directors might also shift a famous legend to a different place. Imagine a vampire from Eastern Europe showing up in a small African village, or a Southeast Asian spirit traveling with immigrants to the United Kingdom. How would local people react? Could the spirit adapt, or would it lose power in a strange land? These questions inspire creativity and encourage viewers to think about how fear moves from one spot to another.

Local Heroes and Resolutions

Sometimes, horror films from certain regions show that local rituals or knowledge can protect against dark forces. A ritual performed by a village elder might calm an angry spirit. A wise person might explain that you have to apologize in a specific way or return stolen objects. These story elements remind viewers that tradition can hold answers.

In other cases, modern science or unity among neighbors saves the day. The main characters might come from different backgrounds, showing that cooperation is key. Horror can reveal that while evil forces exist, people can fight back by combining old and new wisdom.

The Future of World Horror

As technology evolves, so do ways of telling scary stories. Many places now have skilled special effects teams. Even if the budget is small, creative directors can use digital tools to produce polished visuals. Some mix these tools with practical effects for realistic results. Also, online platforms give wide exposure to short films or web series, so new directors can experiment without needing large funding.

Horror also intersects with other genres. You might see a crime drama with supernatural hints, or a romance tested by a ghostly presence. This blurring of genres leads to unique stories that do not fit in a single box. Regions that once produced only a few horror films are now exploring it more, especially when they realize how big the global audience is.

Conclusion

Horror around the world is as diverse as the lands and cultures that produce it. In East Asia, pale ghosts and long-haired spirits reflect beliefs about the dead and the afterlife. In Southeast Asia, curses and vengeful spirits roam forests and small villages. In South Asia, people grapple with black magic or angry ghosts shaped by tradition. The Middle East weaves stories of jinn or ancient curses. Africa may show witchcraft or local creatures, while Europe carries on gothic styles and modern psychodramas. Latin America blends colonial history with local myths, and North America offers Hollywood's wide mix alongside Canadian twists and Mexican legends. Australia and New Zealand add the harsh Outback or secluded mountains to the list of haunting settings.

Each place adapts horror to its own myths, social problems, and laws. Yet viewers all over the globe can relate to that deep, uneasy feeling when the unknown stirs. Different languages and symbols appear, but the fear remains universal. Thanks to modern media, these styles cross borders, influence each other, and lead to new tales that combine old beliefs and modern concerns.

CHAPTER 10: THE PART OF MUSIC

When you watch a horror movie, you might notice that the images alone are not always what scare you. The sounds—from sudden bangs to eerie music—are just as important. In many cases, music sets the mood before you even see something frightening. A single note, played at the right time, can make your heart race. An absence of sound can also heighten tension, making every small noise much more alarming. This chapter explores how music and sound design shape the horror experience.

Why Music Matters in Horror

Music is a powerful tool because it can send signals to our emotions. Even if we do not know the technical details of chords or harmonies, our brain reacts to them. A slow, low-pitched note can feel menacing. A high, screeching violin can make us tense. Horror directors use this knowledge to influence how we feel about a scene.

In many scary moments, the music builds up. You hear a series of repeating notes that speed up or get louder. This causes your heart to beat faster in anticipation. When the final scare arrives, the music might crash into a loud chord. Even if we expect it, the sound can make us jump. Our minds are wired to respond to abrupt changes in volume or pitch.

Silent Era and Early Film Scores

In the earliest horror films, such as *Nosferatu* (1922) or *The Cabinet of Dr. Caligari* (1920), there was no recorded dialogue or sound effects. Instead, live musicians (often a pianist or an organist) played in the theater. They followed a score that matched the moods on screen. They might play eerie chords during spooky parts and switch to dramatic runs when a character was chased.

Because these films had no spoken words, the music was vital. It carried the emotion of each scene. While technology was limited, the principle still applies to modern horror. Music clues us in that something frightening may happen soon. Without it, a scene might feel flat.

Classic Hollywood Horror Scores

When sound entered cinema, studios hired composers to create full music tracks for their films. Universal's classic monster movies (like *Dracula* or *Frankenstein*) had orchestral scores that emphasized the gothic feel. Dark, heavy tones on the organ or strings gave a sense of doom. Sweeping melodies hinted at tragic romance.

In the 1950s, horror leaned into science-fiction themes. Composers used electronic sounds or unusual instruments to suggest aliens or giant creatures. This was the start of exploring new audio effects to make viewers uneasy. The music sometimes used beeping gadgets or dissonant chords that were not common in other genres.

Establishing Themes for Monsters

One common technique is to give a monster or villain a musical motif—a short theme that plays whenever they appear or are about to strike. This can be a sequence of notes that becomes tied to the character. For example, John Williams's famous two-note motif in *Jaws* signaled the shark's presence. Hearing those notes alone can bring the film to mind.

This approach works because our brains link the music to the threat. Once we hear the familiar tune, we anticipate the danger. Even if the villain does not appear right away, the music sets us on edge. It becomes a kind of warning within the film, but it can also act as a surprise if it is used in an unexpected way.

The Rise of Electronic and Synth Music

In the 1970s and 1980s, synthesizers became more affordable and popular. Horror directors embraced them for their strange, sometimes futuristic sounds. John Carpenter used a simple synth score in *Halloween* that is still recognized worldwide. The repeating piano melody, combined with a steady electronic beat, created tension with very few notes.

Other films from this time used synth music to create unsettling atmospheres. The instruments could produce eerie humming, electronic howls, or pulsing beats that seemed unearthly. Audiences were used to rich orchestras in classic cinema, so these new electric tones felt harsh and unnatural—perfect for horror.

Iconic Horror Composers

Several composers have made names for themselves in horror, developing styles that shape the genre.

Bernard Herrmann

Though not solely a horror composer, Herrmann's work with Alfred Hitchcock set the template for how music can enhance suspense. His score for *Psycho* (1960), particularly the shrieking violins in the shower scene, is legendary. The high-pitched strings hit the audience like a scream. Herrmann often wrote music that stayed in a narrow range of notes, repeating patterns to create anxiety.

Jerry Goldsmith

He composed music for *The Omen* (1976), using choral arrangements that sounded like ominous chants. The Latin words, sung in a menacing way, added a feeling of doom and religious dread. This set an example of how a choir can make a film's threat seem larger than life.

John Carpenter

Also the director of *Halloween*, Carpenter composed many of his film scores. He preferred simple but effective melodies, often played on a synthesizer or piano. His music for *The Fog* and *Prince of Darkness* also used minimal motifs, letting the repetition pull viewers into a tense state.

Goblin

This Italian progressive rock band is known for their collaborations with director Dario Argento on films like *Suspiria*. They mixed rock instruments, electronic elements, and choral voices to create a dreamlike, sometimes chaotic sound. The music was as bold as the visuals, turning the film into a total sensory experience.

Christopher Young

He scored films like *Hellraiser*, using dark, orchestral sounds with heavy percussion and choral sections. The result was a grand, hellish atmosphere. Later, he worked on many horror and thriller projects, adapting his style to fit different moods.

Joseph Bishara

Bishara is known for his work with James Wan on films like *The Conjuring* and *Insidious*. He often blends traditional instruments with disturbing sound design, resulting in sudden bursts of noise or odd metallic scrapes. This keeps viewers unsettled, not sure when the next jolt will come.

These composers show that there is no single style that defines horror music. Some prefer elegant orchestras, others weird synth sounds, and others a mix of choral chanting and distorted noises. The unifying goal is to provoke fear or suspense.

Jump Scare Sounds

Jump scares are moments when something suddenly bursts onto the screen or a loud noise happens unexpectedly. The music or sound effect is critical to making this work. Often, we hear a sudden spike in volume—like a screech of strings, a crash, or a harsh electronic burst.

A well-done jump scare might be set up by quiet music or silence right before. This contrast makes the loud sound even more startling. If jump scares happen too often, though, viewers might see them coming. Good directors and composers pace them carefully, ensuring each lands with force.

Subtle Underscoring

Not all horror relies on loud sounds. Sometimes, a faint hum or low drone in the background can work wonders. This approach sets an uneasy tone that builds over time. It can be a quiet heartbeat-like thump that slowly increases, or a near-inaudible chant that suggests the presence of something unseen.

This technique plays on our tendency to sense changes in our environment. Even if we do not consciously notice the rising pitch, we feel the tension. By the time the scary moment arrives, we are already on edge. Subtle underscoring can be more unnerving than a full orchestra if used well.

Silence as a Tool

Silence might be one of the strongest tools in horror. When the music stops abruptly, we focus on every small sound—footsteps, breathing, or a doorknob turning. Our mind prepares for something to shatter that silence. In some cases, nothing happens, which builds dread even more.

A completely silent scene draws us in because we listen closely for clues. Directors often use this during a tense buildup, right before a dramatic event. The viewer thinks, "Something must be coming," creating self-imposed anxiety. If a loud noise suddenly breaks the silence, it can have a huge impact.

Sound Design Beyond Music

Apart from musical scores, horror depends on sound design for effects like creaking doors, howling winds, or squishy gore noises. These sounds can be as scary as any melody. Sound designers collect or create these effects, sometimes recording odd methods—like twisting vegetables to mimic bone cracks.

A well-crafted horror film layers these sounds so that they appear realistic or suitably exaggerated. A hallway might echo with dripping water, or a forest might crackle with twigs snapping. When combined with the music, these details put the audience in the character's shoes, hearing what they hear.

Cultural Instruments and Themes

In global horror, composers might use instruments traditional to a region. A Japanese horror movie could include the shakuhachi (a bamboo flute) or taiko drums. A Middle Eastern film might have the oud or special vocal styles. This adds authenticity and can set the film apart from Hollywood norms.

The choice of instrument can also tie to the story's setting. If the film is about an old curse, the music might include ancient-sounding instruments to hint at the past. If it is a modern story with technology, the music might have electronic elements. Matching the score to the film's theme gives it a stronger identity.

Using Music in Character Themes

Some horror films give each main character a musical theme. When that character faces danger, their theme might twist into a scarier version. If a character becomes possessed, the once-innocent melody might become discordant. This approach is common in many film genres, but in horror, it highlights the shift from safety to terror.

For example, if we have a mother worried about her child, her gentle lullaby theme might appear early in the film. Later, if a demon targets the child, that same melody could reappear in a minor key or with screeching strings, indicating that the mother's comfort has turned into horror. This contrast can deeply affect viewers.

Ambient Horror Scores

Some composers create ambient tracks—long, droning sounds with minimal structure. This approach aims to surround the audience in an eerie atmosphere without a clear melody. Ambient scores can feel relentless, as if the film is coated in a layer of tension.

Movies that focus on slow dread might use this style to avoid overshadowing the visuals. The audience picks up on the mood, but it is never quite a "tune" they can hum. Ambient music works well in psychological horror, where the goal is to unnerve viewers rather than shock them with loud stings.

When Music Contrasts the Scene

Sometimes, horror directors use cheerful or nostalgic music to contrast a terrifying scene, creating a feeling of disorientation. For instance, imagine a bright 1950s pop song playing while a character runs from a masked killer. The mismatch can make the scene even more disturbing, because it feels wrong.

This trick can highlight the brutality of the moment by playing something that does not fit. The audience wonders why a happy tune is there. The effect can be darkly ironic, suggesting that the killer or the environment does not care about the horror taking place. It can also become iconic if the song is memorable.

How Viewers React Physically

Music in horror triggers physical responses. A sudden, loud chord can make us jump in our seats. A low-frequency rumble can cause our stomachs to tighten without us knowing why. Certain pitches or repetitive notes can mirror the body's own rhythms, like a heartbeat, increasing our pulse in sync with the soundtrack.

Filmmakers and composers take advantage of this. By controlling volume, tempo, and frequency, they guide our reactions. This is why horror music can be so effective—it taps into basic human responses to sound. We sense danger when we hear a loud bang, even if we consciously know it is from a movie.

Memorable Title Themes

Many horror films have opening title music that sets the stage. A slow build of eerie notes might play over images of dark hallways, old photos, or cryptic symbols. By the time the film starts, viewers already feel uneasy.

Sometimes, the title theme becomes famous on its own. Fans might recognize it instantly and recall the fear they felt watching the movie. This theme can reappear throughout the film at key moments, tying everything together. When the end credits roll, hearing that theme again can bring closure—or remind viewers they just experienced a scary story.

Avoiding Clichés

While music is crucial, it can be overused. Some horror films rely on loud
stingers for every small event, which can wear out the audience. When
everything has the same musical shock, it loses effectiveness. Good composers
and directors balance quiet scenes with musical ones, building a proper rhythm
of tension and relief.

Overly dramatic music can also be a problem if it does not match the tone. For a
subtle, psychological story, a bombastic orchestra might feel out of place. For a
supernatural tale, a gentle acoustic guitar might not capture the sense of terror.
The music must fit the film's identity, or it risks pulling viewers out of the
experience.

Live Instruments vs. Digital

Some composers still prefer recording live orchestras. The emotional range of
real strings or horns can be richer than a digital sample. Others lean on digital
libraries and synths because they can craft unusual, creepy sounds that are hard
to achieve with acoustic instruments. Many blend both approaches.

Budget plays a role as well. A small production might not afford a full orchestra,
so a composer uses digital tools. Still, creativity can shine through on any
budget. A single cello, played in eerie ways, might be enough to unsettle viewers
if used cleverly.

Songs with Lyrics

Sometimes, horror films feature songs with words, either existing tracks or
written for the film. Lyrics can foreshadow the plot or hint at the villain's motive.
If a song references lonely roads or drowning sorrow, it might fit a movie about a
haunted lake.

A well-placed song can also become iconic. For example, if a certain tune plays
whenever the killer is near, viewers will associate that song with danger. This
might extend outside the movie. Hearing it on the radio could remind people of
the film and bring back those tense feelings.

Music During the Climax

As the story reaches its peak—when characters confront the threat—music often ramps up. The tempo might increase, the notes might get higher, or the volume may swell. This mirrors the characters' desperation. If they succeed, the music may shift to a triumphant or peaceful tone. If they fail, the music might drop to a haunting quiet or a final blast of tragic sound.

Some horror films end with a sudden silence after the climax, leaving viewers shocked. Others maintain unsettling chords through the credits, suggesting the evil still lingers. The composer and director decide how they want the audience to feel when the story ends—relieved, haunted, or unsure.

Viewer Listening Habits

Outside the theater, fans might listen to horror soundtracks to recapture the film's mood. Some find it relaxing, others find it unsettling. A soft, creepy track can be good for reading spooky stories, while an intense orchestral piece might be too stressful for casual listening.

Streaming platforms have playlists titled "Scary Music" or "Horror Soundtracks," showing there is a market for these scores. People might play them at events with haunted themes or just to evoke a certain atmosphere at home. The music, separate from the images, still holds power because it triggers memories of the film's scary moments.

Evolving Trends

As technology evolves, composers experiment with new sounds. Some use digital distortion to create audio illusions, so the viewer cannot tell where the noise comes from. Others integrate real-world recordings—like city traffic, rustling leaves, or animal calls—into the score to blur the line between the movie's world and reality.

Virtual reality horror experiences go even further. The music and sounds react to the user's actions. If someone turns their head, the audio changes direction.

This level of interaction intensifies fear, because the viewer feels more immersed. Composers must think differently for these formats, designing adaptive music that shifts based on real-time movement.

Conclusion

Music is a key part of what makes horror films frightening. It can warn us before we see the threat, or it can surprise us with a sudden crash. Composers use instruments, electronic tones, voices, and even silence to shape our emotions. We might not always notice how carefully the sounds are planned, but we feel their impact.

Through history, horror music has grown from simple live piano to complex orchestral scores, from eerie rock band tunes to digital drones. Each style can be effective if it matches the film's tone. The best horror scores do more than support the visuals—they become part of the film's identity. When we think of that movie, we can almost hear its notes echoing in our mind.

By paying attention to the music next time you watch a horror film, you might notice moments where a subtle chord change raises your heartbeat. You might spot how silence makes your skin crawl. You might even find a theme that sticks in your head long after the end credits. All these elements show how deeply music affects our experience of fear and suspense. Whether it is a haunting lullaby or a thundering orchestral blast, music is the unseen guide that leads us into darkness—and sometimes follows us back out with a final, chilling note.

CHAPTER 11: THE ROLE OF SPECIAL EFFECTS

Special effects are one of the main ways horror films show terrifying sights. They can bring monsters to life, create shocking injuries, or turn ordinary settings into nightmarish worlds. Good special effects help viewers believe in what they see. Even if they know it is not real, a convincing effect can still make them feel frightened or uneasy. Over the years, special effects have grown from simple tricks to advanced digital creations. In this chapter, we will look at different ways filmmakers use special effects to shape horror, including makeup, animatronics, computer-generated imagery, and more.

Early Techniques

In the earliest horror films, there was no modern technology for special effects. Filmmakers had to be resourceful. They used camera tricks, basic props, and editing to make strange things happen on screen. For example, a common trick was stopping the camera to replace an actor with a dummy, then starting the camera again. This "stop trick" made it look like a person vanished or changed shape instantly. Audiences at the time found it startling because they had never seen such illusions.

Silent films also relied on makeup and lighting to make monsters look scary. Filmmakers painted shadows on walls or used mirrors to create eerie reflections. Actors wore heavy greasepaint to highlight their expressions. Though it was low-tech, it set the stage for more complex work in later decades. Even these basic methods showed that horror relies on visuals that disturb or surprise viewers.

Practical Makeup Effects

One of the oldest and most trusted forms of special effects is makeup. Makeup artists spend many hours applying products to an actor's face and body, turning them into zombies, vampires, or other creatures. This can range from a little bit of fake blood to a full head-to-toe creature suit. Skilled makeup artists learn about paints, prosthetics, and molds, so they can craft realistic wounds or monstrous features.

Prosthetics

Prosthetics are fake pieces attached to the actor's skin. They can be small, like scars or pointed ears, or large, like an entire werewolf face. To make them, artists first sculpt the desired shapes in clay. Then they use materials like latex or silicone to create soft, lifelike pieces. These pieces are glued onto the actor, and the edges are blended with makeup. The final result can look very real, especially under good lighting.

Foam Latex and Silicone

Older horror films often used foam latex, which is lightweight and flexible. When painted and glued on, it moves with the actor's facial muscles. Modern artists also use silicone, which can look more like actual skin. Silicone pieces have a translucent quality, letting light pass through them in a way that feels natural. This helps sell the illusion that the actor truly has monstrous features or open wounds.

Coloring and Texture

A key part of makeup effects is painting. Makeup artists use many layers of color to simulate bruises, veins, or rotten flesh. They also add texture, like wrinkles or scaly patterns. This takes time and care, since a rushed paint job can look fake on camera. The best artists study real injuries or animal skin to understand how to mimic them. The more detail, the more believable.

Famous Makeup Artists

Several makeup artists became well-known in the horror field. For instance, Jack Pierce created the iconic look of the Frankenstein monster for Universal's 1931 film. He used cotton, collodion, and spirit gum to build up the monster's flat head and scars. Decades later, artists like Tom Savini impressed viewers with realistic gore effects in zombie and slasher movies. Savini's experience in practical gore techniques set the standard for many horror films in the 1970s and 1980s. More recently, people like Stan Winston's studio and Greg Nicotero's team have crafted memorable creatures, from dinosaurs in other genres to walkers in zombie series. Their work shows that skilled hands-on effects can sometimes look more real than digital creations.

Animatronics and Puppetry

In addition to makeup, many horror productions use animatronics and puppets
for creatures. An animatronic is a mechanical figure controlled by motors and
cables. Puppetry might involve sticks, strings, or a performer inside the creature.
Both methods let the beast move in front of the camera in real time. Actors can
interact with it, which can help their performances seem more genuine.

Building the Machine
Creating an animatronic monster is a major project. Technicians design the
metal or fiberglass frame, install motors for movement, and connect them to
controls. On top, they place foam latex or silicone "skin" to hide the machinery.
Skilled painters add details like fur, scales, or slime. The final piece could be a
large head with moving jaws, a tail that swings, or even a full-body creature that
can stand up.

Real Reactions
A big advantage of animatronics is that actors can see the creature on set. They
can touch it, watch it breathe, or see it snap its jaws. This often leads to a more
believable scene than if they must imagine a monster that will be added later
with digital work. The crew can adjust the puppet's movements in real time,
making small tweaks to the performance. Animatronics can also cast real
shadows, interact with lighting, and feel present in a way that purely digital
creatures sometimes struggle to match.

Challenges and Limits
Animatronics can be expensive and hard to operate. A single creature might
need several operators. One controls the mouth, another the eyes, another the
arms. If something breaks, production might halt until it is fixed. Also,
mechanical puppets can be heavy or unwieldy. They may not move as fluidly as a
digital character. Directors must plan shots carefully so the illusions stay strong.
But even with these hurdles, animatronics remain a favorite among many horror
fans who appreciate the physical presence on screen.

Old-School Tricks: Miniatures and Matte Paintings

Before computer graphics, filmmakers used miniatures and matte paintings to
depict sets or large-scale scenes. For instance, if a giant creature was destroying

a city, the crew might build a small model of that city with tiny buildings. They then filmed the monster—sometimes a person in a suit—stomping through the miniature. When done well, this technique could look quite real.

Miniatures

Houses, cars, even trees can be scaled down to create illusions of big environments. If the camera angles are right, viewers may not notice they are looking at a table-sized set. The same approach can be used for big objects like spaceships or huge monsters. When combined with slow-motion filming, miniature sets can mimic real-life physics, because small models tend to move differently than large ones.

Matte Paintings

Matte paintings are detailed artworks that extend a set beyond what is physically built. For example, if a director wants a castle on a cliff, they might film the actors in a partial set. The rest is a painting that lines up with the live part. Today, matte paintings can be digital, but old films used physical paintings on glass. Artists matched perspective and lighting to blend the painting with the real footage. Horror films used this to show spooky castles, giant caves, or endless graveyards without building them all.

These tricks required patience and skill. The perspective had to be perfect, or the illusion would break. Even so, they gave horror filmmakers a way to create huge or fantastic environments when budgets were small.

The Rise of Digital Effects (CGI)

Starting in the late 20th century, computer-generated imagery (CGI) became a big part of movie making. Horror was no exception. CGI allowed creators to show things that would be impossible or too expensive with practical methods. Ghosts could float through walls, giant creatures could run across the screen, and transformations could happen smoothly without cuts.

Pros of CGI

1. **Flexibility:** A digital creature can move in ways a puppet cannot. You can make it fly, stretch, or shapeshift.

2. **Speed**: Once the digital model is built, artists can change or refine the animation on a computer without reconstructing props.
3. **Safety**: Certain stunts or gore effects can be done digitally, reducing risk to actors.

Cons of CGI

1. **Costs**: High-quality CGI can be very expensive. It requires skilled artists and powerful computers.
2. **Risk of Looking Fake**: Poorly done CGI might look cartoonish. If viewers do not believe the creature is real, the scare factor drops.
3. **Lack of Physical Presence**: Actors may have to react to green screens or placeholders. This can weaken performances if not handled well.

Over time, CGI has improved. Some horror films combine both digital and practical techniques for the best outcome. For example, a director might use an animatronic for close-ups, but switch to CGI for wide shots of the monster running. This balance can keep the realism while allowing dynamic action.

Blending Practical and Digital

Many horror creators find that blending practical elements with digital enhancements yields the most convincing results. For instance, an actor might wear a partial creature suit, with the head and arms physically there. Then the effects team uses CGI to add extra movement, glowing eyes, or bigger claws. This helps maintain a sense of weight and tangibility while letting the effects team polish details.

In gore effects, makeup might show the main wound, but digital artists remove safety straps or add dripping blood that flows in precise ways. Or if a scene needs an impossible transformation, the early phases are done with prosthetics, then morph into a CGI transition, and finally return to practical effects when the new form is complete. The viewer does not always notice these switches if they are seamless.

Iconic Creature Suits

Before CGI became common, creature suits were a staple of horror. An actor or stunt performer would wear a bulky costume, sometimes with built-in animatronics for the head. Classic examples include the Gill-man suit from *Creature from the Black Lagoon* or Godzilla suits from Japanese kaiju films. While modern audiences might find these suits less realistic compared to advanced effects, many fans appreciate their charm and the genuine performance an actor could give inside the suit.

Suit Acting

Performers inside suits had to convey personality through limited facial movement. They used body language, posture, and gesture. With thick rubber or latex covering them, it was hot and difficult work. They had to be careful not to tear the suit or faint from the heat. Nevertheless, when done well, suit acting added a layer of authenticity, because the monster physically existed on set.

Suit Modifications

Crews sometimes added fins, spikes, or large tails. They might hide tubes that pump out slime or other fluids. The eyes could be controlled by a puppeteer off camera. In some cases, suit actors had to walk on stilts or adopt a crouch to achieve a certain shape. All these details combined to make the monster stand out and feel memorable.

Blood, Gore, and Shock

Gory effects are common in certain horror subgenres, like slashers or zombie films. Whether it is a splatter of blood or a severed limb, these visuals can shock audiences. The realism depends on the materials used.

Fake Blood

Classic fake blood recipes mix corn syrup, food coloring, and sometimes cocoa powder for darker shades. In modern times, professional labs produce blood with different thicknesses or colors. Thin, bright blood might look fresh; thick, dark blood might suggest old wounds. Directors might demand gallons of fake blood for big scenes. Getting the right color on camera is tricky. Under certain lights, fake blood might appear too bright or too dull. Experts test it in the same lighting conditions the scene will use.

Latex and Silicone for Gore
To simulate injuries, makeup artists create prosthetics for wounds, exposed
bones, or torn flesh. These pieces might have tubes inside to pump fake blood.
Actors can wear them like patches on their skin. When the scene calls for a cut,
the director signals the blood pumps, creating the illusion of a fresh wound. If
the camera lingers, the makeup must hold up to close inspection.

Body Parts
For severed limbs or heads, the team might create life-size replicas. They use
molds of the actor's body, then paint the cast with lifelike detail. Hair punching
(adding individual hairs) can make a severed head look very convincing. Some
horror films store entire closets of spare arms, legs, and heads to prepare for
messy scenes.

Transformations

A highlight in many horror stories is a character who transforms, like a human
turning into a werewolf. Showing this change can be challenging. In older films,
creators used dissolves between different makeup stages. The actor might freeze
in place, the camera would pause, makeup artists would swap in new
prosthetics, then the scene would continue. When edited together, it seemed
like the transformation happened on screen.

Later, mechanical heads allowed slow changes. In *An American Werewolf in
London*, the transformation sequence used stretching latex pieces, hair growing
through holes, and animatronic extensions. It took multiple days to shoot, but
the result stunned audiences. Modern films can use CGI to refine the process
further, yet many fans still praise the ingenuity of those mechanical
transformations.

Ghostly Effects

Not all horror focuses on blood or monsters. Some emphasize ghosts and spirits.
Creating ghostly visuals can involve layered images, transparent overlays, or
digital compositing. For example, if a director wants a faint silhouette to drift

through a wall, they can film the actor against a green screen, then layer that footage over the house set. Adjusting opacity makes the figure partially see-through.

In older times, double exposure was used. Filmmakers shot a scene on the same strip of film twice. First, they filmed the empty hallway, masking part of the lens. Then they rewound the film and shot the actor in the masked area. When developed, the actor looked transparent. Such tricks gave early ghost films a spooky feel. Modern digital methods offer more flexibility, but the idea remains the same: showing something that is there and not there at once.

Fire, Explosions, and Destruction

Though not unique to horror, big scenes with fire or destruction add tension. When a haunted house goes up in flames at the climax, it can be both dramatic and dangerous. Practical fire stunts rely on trained performers wearing protective gear. Sets might use controlled gas lines to manage flames. If things get out of hand, firefighters on set can intervene. Some directors choose digital flames for safety, but realistic digital fire requires skill to look believable.

Explosions might appear if characters destroy a monster's lair or blow up a lab gone wrong. Practical explosions use carefully set charges. A mixture of dust, debris, and flame can make it look huge. Meanwhile, cameras capture the blast from multiple angles. In horror, a well-timed explosion can mark the end of the threat—or reveal something worse survived.

Miniature and Model Work Today

Even with CGI, some horror productions still use miniatures. They might build a detailed model of a spooky mansion's exterior to film certain shots. Later, they add digital effects like swirling clouds or lightning. This approach can be cheaper than building a large set, and it can look more authentic than a fully CG environment. By mixing physical models with digital enhancements, creators get both realism and flexibility.

Challenges of On-Set Effects

Practical effects can go wrong if not planned carefully. A fake blood pump might fail, leaving a scene half-done. An animatronic might break a motor or lose power. If that happens at a critical moment, production halts. A transformation scene might require hours of makeup, so the actor must remain still, risking discomfort. All these factors add stress to a shoot, but many directors feel the payoff is worth it. Audiences often appreciate the tactile quality of real objects on screen.

Directors also have to account for extra time in the schedule. Complex makeup might need four hours each day before filming. If the shoot runs late, the actor endures many hours in a heavy costume. This can strain the team, but when the final footage captures a terrifying monster, it can be a highlight of the film.

Computer Tools for Editing Effects

Even when practical effects are used, digital editing programs can fine-tune details. Artists might remove wires or rods that control a puppet. They might erase safety equipment around a stunt performer. If an animatronic's movement seems jerky, digital smoothing can fix it. Or if the makeup edges are visible, they can blend them out in post-production. This synergy between old-school craftsmanship and new software helps horror filmmakers achieve polished results.

Sound Effects as Part of the Visual Illusion

It is easy to think only of visuals, but sound is vital in selling special effects. A bloody slash on screen is more intense if we hear a sharp swish and a wet impact. A giant creature's roar matters as much as its appearance. Even a ghost effect seems scarier with a distant moan or rattling chain. The mix of audio and visuals can convince our brains that the threat is real.

When a prosthetic head is bitten off, for instance, the scene might cut quickly while playing a crunch sound. Our minds fill in the gaps. We do not always need to see everything; the suggestion, aided by sound, is enough. By pairing a carefully built effect with matching audio, horror filmmakers engage multiple senses at once.

Trends in Modern Horror Effects

Lately, many horror filmmakers have returned to practical effects, even as CGI remains common. Some say that well-made animatronics and makeup are timeless. Others prefer the freedom of digital effects, especially for supernatural elements that require improbable motion. The best horror often uses both. For instance, a film might have actors in real creature suits, but glowing eyes and smoke-like auras are added digitally.

Television shows and streaming productions also invest heavily in horror effects. With multi-episode arcs, teams can refine creature designs across a season. Viewers might see subtle changes as the monster evolves, or bigger reveals in later episodes. This long format allows more character interaction with the effects, making them feel integrated into the story.

The Magic of Believability

No matter how advanced the technology, the main aim of special effects in horror is to help viewers suspend disbelief. If the audience sees a rubber suit or notices obvious digital seams, the scare might not land. But if everything blends smoothly, viewers will react with fear or disgust, even while knowing it is fiction. The artistry lies in that balance between tricking the eye and leaving room for imagination.

Directors and effects teams often emphasize restraint. Showing the monster in small glimpses can be scarier than full exposure. If the creature is fully revealed, the design must be strong enough to maintain tension. Sometimes, the best approach is to let shadows and partial views do the work, with only a brief highlight of the detailed makeup or digital model. The viewer's mind fills in the rest, which can be more chilling than any effect on screen.

Fan Appreciation and Behind-the-Scenes

Many horror fans enjoy learning how these effects are made. Documentaries and behind-the-scenes clips show the hours of makeup, the engineers operating puppet rigs, or the computer artists animating ghouls. People who watch these features often gain respect for the time and effort poured into each effect.

This interest leads to conventions where makeup artists and effects wizards show off their work. They might do live demos, applying prosthetics to volunteers or showing off animatronic heads that blink and snarl. Fans ask questions about materials and methods. Such events reveal the care behind each gory detail, reminding us that the horror genre involves creativity and craft.

The Future of Horror Effects

Technology keeps evolving. Virtual reality (VR) horror experiences can layer real-time digital effects over a user's view, making them feel immersed in a haunting. Augmented reality (AR) could let watchers see ghosts appear in their living rooms. These new frontiers might push the boundaries of how we experience scary stories. But even as new tools arise, classic practical effects are unlikely to vanish. Many filmmakers see them as part of horror's soul. The hands-on approach fosters a tangible sense of reality that purely digital images sometimes lack.

We may also see more advanced motion capture, where actors in suits covered with tracking dots perform creature movements. Computers then render a beast that mimics the actor's motions. This method captures subtle body language that animators might not replicate by hand. The result can be a monster that moves in very human, yet distorted, ways—perfect for unsettling an audience.

CHAPTER 12: MORALS IN HORROR

Horror is best known for its scares, but it also carries moral themes. Many horror stories show people facing the results of their actions, or they use monsters to symbolize deeper lessons. Even a violent slasher film can have a moral side, teaching viewers about choices and consequences. In this chapter, we will explore how horror films handle morals, ethics, and the idea of right and wrong. We will also look at the ways these stories reflect real issues in society.

Roots in Cautionary Tales

Long before movies, people told scary stories around fires. Often, these tales warned listeners about bad behavior. If someone wandered off alone at night, a monster might snatch them. If someone broke a rule, a curse could follow. These cautionary tales aimed to keep communities safe by using fear to enforce moral conduct. Horror films inherit this tradition. Though modern stories might look different, the idea of punishment for wrongdoing remains common.

The Idea of Punishment

In many horror plots, characters who act rudely, selfishly, or harm others often get the worst fates. This is sometimes called the "moral code" of horror. The film may not state it directly, but viewers notice it. For instance, in certain slasher movies, the character who bullies or lies might be the first to fall victim. While some think it is a cliché, others see it as a reminder that bad actions can catch up to you.

Reinforcing Societal Norms

When a film shows that cruelty leads to a grisly end, it can send a message that being kinder or more respectful is better. Of course, not all horror uses this approach. Some movies show random violence, where good and bad characters suffer equally. Still, the concept of punishment for wrongdoing appears often enough that many fans think of it as a genre staple.

Karma in Action

In some horror stories, the idea of karma is clear. A person who steals might be haunted by the victim's ghost. Someone who abuses nature might face a monster that represents the Earth's anger. The direct link between the action (stealing or harming nature) and the result (a supernatural attack) forms a moral lesson: if you harm others, consequences can follow.

Facing Personal Guilt

Another moral angle in horror involves characters dealing with guilt. A person might have caused someone's death by accident or through neglect. Later, they see ghosts or nightmares that reflect their remorse. The haunting is not random; it is tied to the character's past deeds. The film's tension comes not just from the supernatural events, but from the character's inner battle. They might only find peace by confessing or making amends.

Examples

- A mother in a horror film might feel guilty about a child she lost. Strange visions or eerie sounds echo that guilt, forcing her to face what happened.
- A driver who fled from an accident might see a phantom figure that keeps appearing, symbolizing the victim's unpaid debt.

In these cases, the horror acts like a judge, forcing the main character to confront what they have done. The moral is that you cannot run from your conscience.

Temptation and Forbidden Knowledge

Some horror stories revolve around forbidden knowledge—like summoning spirits or reading cursed books. Characters who seek power or insight beyond safe limits might unleash dangerous forces. The moral suggests that greed or curiosity can lead to ruin if not guided by caution.

Dealing with the Unknown

Films about ancient curses often show explorers ignoring warnings. They break seals, enter tombs, or read texts in strange languages. Soon, a terrifying entity appears. The moral is that certain things are better left alone. This echoes older myths about mortals who anger the gods by crossing boundaries. In modern versions, the "gods" might be malevolent spirits or alien beings, but the lesson stays the same: arrogant acts can bring downfall.

Morals About Society

Some horror films point out societal problems. They might use monsters or dystopian scenarios to show how humans treat each other under stress. If characters turn on each other instead of uniting, the monster has an easier time picking them off. The moral might be that division or cruelty is more dangerous than any external threat.

Groups in Crisis

In many zombie films, survivors hide in a building. Instead of cooperating, they argue and split into factions. This leads to chaos. The real horror is not always the zombies outside, but the humans inside who cannot work together. A moral emerges: unity and empathy are crucial in dire times. Without them, people invite disaster.

Social Critique

Some horror uses extreme scenarios to shine a light on injustice. A film might show a wealthy family ignoring the needs of the poor, only to be haunted by ghosts that represent past wrongs. Or it might depict a corporation doing shady experiments that unleash a terrible creature. The moral is that greed or lack of compassion can doom everyone. Horror amplifies these issues to make the message hit harder.

Symbolic Monsters

Monsters themselves can be symbols. A vampire might stand for exploitation, taking from others to survive. A werewolf might symbolize rage that a person cannot control. Ghosts might represent painful memories that refuse to be forgotten. By personifying fears and failings, horror gives a form to moral conflicts. Characters who defeat the monster might learn to overcome their own flaws. Or they might fail, showing how deadly those flaws can be.

Dr. Jekyll and Mr. Hyde

Though more of a dark tale than a straight horror film, this story shows a man who splits his good and evil sides. The moral is that everyone has a hidden darkness. If we let it run free, it can destroy our lives. Many horror stories borrow this theme, presenting a monster that is really a reflection of the main character's worst impulses.

Final Girl and Moral Codes in Slashers

A well-known idea in slasher films is the "final girl." This is typically a character who survives until the end. In older films, she is often shown as more responsible or reserved than her friends. Some fans argue that these slashers push moral standards—suggesting the final girl lives because she avoids certain reckless behaviors. Others say this reading is too simple and that the final girl is just an archetype to build tension.

Changing Times

In modern slashers, the final girl trope might be subverted. Perhaps she fights back not because she is innocent, but because she gains the strength to face the killer. Or maybe the film shows multiple survivors, not only the "nice" one. Still, the moral idea that wise choices or caution can help you survive remains part of many horror stories.

Exploration of Moral Gray Areas

Not all horror is black and white. Some films present choices where there is no easy right answer. Characters might do something questionable to save themselves, and the story asks whether that is justified. This can lead viewers to question their own moral lines.

Examples

- A group is trapped in a location with limited supplies. One character suggests sacrificing someone to keep the rest alive. Is this survival or murder?
- A parent might strike a deal with a dark force to heal their child, but it means someone else will be harmed. Is it understandable or evil?

These films show that horror can be more than simple punishments for wrongdoers. It can highlight complex moral dilemmas.

Redemption Arcs

Sometimes, a horror story lets a flawed character redeem themselves. They might start off as cruel or cowardly, then realize they must protect others. By the end, they show courage or compassion, possibly breaking a curse or defeating a demon. The moral is that it is never too late to change. Horror settings heighten the stakes, making redemption feel urgent and dramatic.

Atonement

If a character caused harm before, they might seek forgiveness by confronting the evil force. This can be more powerful than just saying "sorry." They prove their change by risking their life to right old wrongs. Horror's high danger makes each action count, reinforcing that atonement is earned through deeds, not words.

Critiquing Blind Faith

Some horror stories show the dangers of blindly trusting leaders or groups. For example, a cult might follow a twisted belief system, performing rituals that harm outsiders. The moral warns viewers about giving up personal judgment. These stories might say, "Think for yourself, or you could be trapped in something dangerous."

Rituals Gone Wrong

When a film features a cult or secret society, it often shows that group's extreme methods. The line between tradition and cruelty can blur. Members might sacrifice people to appease a deity or secure power. Horror reveals how easily moral values can twist if no one questions authority. The hero typically breaks free by rejecting the cult's rules, implying we should not follow others blindly.

The Role of Fear in Moral Lessons

Fear can make people act in ways they would not otherwise. Some horror films highlight how fear pushes characters to lie, betray friends, or commit violence. The moral might be that panic and paranoia are as much a threat as the monster. If characters had stayed calm, they might have found a better path.

In *The Mist*, for instance, people trapped in a store face strange creatures outside. Fear leads some to embrace extremist thinking. They turn on each other, making the situation worse. While the creatures are terrifying, the moral is that hopelessness and fear can break our moral compass faster than any monster. This reflection on human behavior under stress is a key part of many horror plots.

Social Issues and Allegory

Horror often works as an allegory. An allegory is a story with a hidden meaning that comments on real-world problems. Directors might use zombies to talk about consumerism, aliens to discuss xenophobia, or ghosts to explore trauma. Viewers sense the message beneath the surface. This approach can be more impactful than a direct lecture, because horror hits emotions strongly.

Race and Class

Modern horror films sometimes address race and class conflict. *Get Out* shows how prejudice can hide behind friendly smiles, turning into something horrifying. The moral is that ignoring or denying bias can lead to real harm. Other films might show that a haunted house belongs to people who were treated unjustly, and their spirits demand respect. In each case, the monstrous or supernatural element stands for society's hidden injustices.

Environment

A film might show a polluted lake that spawns a creature, or a forest spirit
punishing those who cut down trees carelessly. The moral is clear: harming
nature can bring dire consequences. Horror dramatizes these issues so viewers
feel the urgency. Instead of a calm argument, it uses fear to drive the point
home.

Moral Ambiguity in Creatures

Some horror creatures are not purely evil. They might simply be following their
nature. This raises questions about whether humans have the right to destroy
them. For instance, a group might kill a beast that was only defending its
territory. Is that heroism or cruelty? The moral can become complex when the
monster's perspective is considered.

Vampire with a Conscience

In a story where a vampire regrets hurting people, the moral might ask if we can
blame a being who must feed on blood to survive. Should humans try to help it
instead? Horror that humanizes monsters forces us to reflect on empathy.
Sometimes, the scariest part is seeing that we can be monstrous in our
judgment.

Teaching Responsibility and Consequences

Many horror films show that careless actions have big consequences. Teenagers
who play with a cursed board game, a scientist who tests an unapproved drug, or
a treasure hunter who ignores warnings: all face supernatural or horrific fallout.
The moral is that ignoring caution can lead to regret. This message can apply to
everyday life, teaching responsibility without sounding like a lecture.

Morals in Ghost Stories

Ghost stories often revolve around the dead who cannot rest. This might happen
because they were wronged, and the living never acknowledged their pain. The

haunting forces characters to learn the truth. By the end, the moral might be about compassion, justice, or the importance of remembering those who suffer.

Sometimes, letting the ghost find peace involves solving a past crime or giving a proper burial. These actions show respect for the dead. The moral is that ignoring problems does not fix them. Only by confronting the past can people move on. Horror transforms this lesson into a chilling experience where the ghost's presence demands attention.

Caution About Blaming Victims

While horror can teach morals, it is important to note that some older films seemed to suggest victims were at fault if they did anything deemed "immoral." Modern viewers often discuss how these portrayals can be unfair or harmful. Not everyone who parties or goes alone somewhere deserves what they get. This conversation reminds us that horror's moral messages can reflect the views of their time, which may need updating.

Hope Amid Horror

Despite the grim tone, many horror films end on a note of hope or resolution. The final survivors might learn a lesson about trust or bravery. They might vow never to repeat the mistakes that caused the crisis. This hints that while bad things happen, people can grow from them. The moral is that facing fears and taking responsibility can lead to a new sense of strength.

Bittersweet Endings

Not all endings are happy. A film might show that the evil was only partially defeated, or the hero loses a friend along the way. Still, if the hero gains moral insight—like regret for past actions or a stronger bond with others—there is a hopeful side. This mix of darkness and growth is common in horror.

Younger Viewers and Lessons

Although many horror films are made for adults, some lighter spooky stories aim at younger audiences. They can deliver moral lessons about sharing, kindness, or facing fears in a less intense way. For instance, a children's animated tale might have friendly ghosts teaching the value of helping others. While not truly scary, these stories show that the idea of moral lessons in horror starts early.

The Debate on Moral Clarity

Some fans enjoy clear-cut moral lessons, such as "crime does not pay" or "treat others well, or face payback." Others prefer ambiguity, where the story raises questions without easy answers. Horror as a genre can cater to both tastes. A slasher might follow classic rules, punishing the wicked and sparing the innocent, while a psychological thriller might leave viewers unsure who was truly at fault. The presence or absence of a strong moral statement often shapes a film's tone.

Cultural Differences

Morals in horror can vary by culture. In one region, respecting elders might be a key lesson, so a spirit targets those who are disrespectful. In another, unity might be the big theme, punishing characters who betray the group. Folks from different backgrounds might find distinct messages in the same film. This range of interpretation makes horror both personal and universal. We all fear something, but the moral spin differs based on local beliefs and social values.

CHAPTER 13: THE PEOPLE WHO LOVE HORROR

Some people cannot stand horror movies. They feel too tense or have nightmares afterward. But others truly enjoy being scared on screen. They seek out frightening films, discuss them with friends, and even collect posters or costumes. This chapter explores who these horror fans are, why they like scary stories, and how they interact with each other. We will look at the social side of horror, the shared excitement among fans, and the ways people handle the intense emotions horror can spark.

Drawn to Fear in a Safe Place

One main reason fans enjoy horror is the safe thrill. They can feel a rush of adrenaline, knowing that there is no real danger. It is similar to riding a roller coaster. You might scream and hold tight, but once it ends, you are fine. Horror lets viewers step into a world of ghosts, monsters, or killers without truly risking their lives.

This sense of safety can be stronger when watching with friends. A group might gather in a living room, turn off the lights, and watch a scary film together. They jump at the same moments, then laugh when the scare is over. This shared thrill helps people bond. Afterward, they talk about the scenes that frightened them most. It creates memories they can revisit later. Some horror fans say the emotional highs and lows bring them closer to their friends.

Different Types of Horror Fans

Not every horror viewer loves the same style. Some prefer classic black-and-white monster movies with gothic sets. Others enjoy modern slashers full of chase scenes. Some like ghost stories with quiet tension. Others like gory zombie flicks. Fans often find a type of horror that fits their personal taste. They might even collect DVDs or stream countless titles in their chosen niche.

1. The Monster Enthusiast
These fans love strange creatures. They study the designs of vampires, werewolves, or made-up beasts. They might draw sketches or create figurines. They enjoy the creativity behind monster designs and are fascinated by the thought of something living among us that breaks normal rules of nature.

2. The Slasher Fan
Slasher fans like suspense and jump scares. They are drawn to masked killers, tense pursuits, and the thrill of not knowing who will survive. Some like the mystery aspect: guessing who might be the villain if the film is set up as a whodunit.

3. The Supernatural Aficionado
Ghosts, demons, haunted houses—these fans want a feeling of dread from the unseen. They like the idea that there could be another realm or presence we cannot fully explain. They often discuss whether they believe in real ghosts or not, and they swap spooky stories.

4. The Psychological Explorer
These fans enjoy horror that focuses on the mind. They like stories about paranoia, guilt, or illusions. They do not need loud scares. Instead, they want to watch characters unravel under psychological stress. They may also enjoy thriller movies that balance mental tension with horror elements.

5. The Gore Collector
Some fans appreciate detailed makeup effects. They watch for realistic wounds, weird creatures, and shocking visuals. They might follow makeup artists on social media and learn how special effects are done. It is not always that they crave violence, but they are impressed by the craft behind fake blood and prosthetics.

Of course, many fans overlap categories. A single person might love old monster classics while also enjoying ghost stories. Horror is broad, giving viewers many styles to explore.

The Social Circle of Horror Fans

Horror fans often form communities—both online and in person. They share reviews, recommend films, and talk about their favorite directors or actors.

These communities can be very passionate. People go to horror conventions where they meet actors from famous films or see displays of iconic props. They might show up dressed as monsters or slashers, wearing carefully crafted costumes.

Online forums and social media groups let fans discuss new releases, debate which monsters are scariest, or ask for hidden gem movie suggestions. Some create fan art or write their own horror stories. Others run podcasts where they analyze horror tropes or interview filmmakers. This sense of community can be a big part of why someone remains a horror enthusiast. They feel understood among people who also like the thrill of being scared.

Collecting Horror Memorabilia

Many horror fans enjoy collecting items related to their favorite films. This can include movie posters, signed photos from actors, or limited edition figurines. Some display these in their homes, turning a corner of the living room into a mini horror museum. Rare or vintage items can become quite valuable. For instance, original posters of classic monster movies might cost a lot, especially if they are in good condition.

Collectors might also gather masks, costumes, or props. They like the feeling of owning a piece of the film's magic. If a movie features a distinct item—like a cursed doll—a replica might become a collector's dream. These objects become conversation pieces, letting fans connect with each other when they spot a rare find.

Why Some People Do Not Like Horror

While there are big horror fans, others avoid this genre. Some viewers have strong emotional reactions or get nightmares. They may feel uneasy for days after watching something scary. People with anxiety or who are sensitive to disturbing images might find horror too intense. They prefer lighter, calmer entertainment.

Personal experiences can shape this too. Someone who faced a real-life traumatic event might not enjoy seeing similar events on screen. The idea of gore or violence might bring back unpleasant memories. This is why some people stay away from horror entirely. They choose to protect their peace of mind.

Halloween and Other Spooky Gatherings

In some places, people love to hold spooky gatherings each year. They wear costumes, watch frightening movies, and decorate their homes with cobwebs or skeletons. Even those who are not hardcore horror fans may have fun during these times, because the mood is playful. They might carve pumpkins, tell ghost stories, or take part in small scare pranks. It shows that horror can be a shared experience, appealing to those who only like mild scares as well as those who crave stronger ones.

In other cultures, different events might revolve around ghosts or spirits. These times can become a chance for fans to gather, share their favorite stories, and watch special film showings. Some local movie theaters run horror marathons at night, playing classic and modern titles for big crowds of fans. People bring snacks, blankets, and a love for jumpy moments on the big screen.

Conventions and Events

Horror conventions happen in large cities around the world. Fans buy tickets to meet actors known for their roles in scary films. They get autographs, take photos, and hear the actors' stories from set. There might be panels with directors or special effects teams who demonstrate how they make creatures or gruesome wounds. Vendors sell collectibles, costumes, and rare DVDs.

These conventions can be festive in nature, with many fans dressing up. You might see someone dressed as a classic vampire, another as a modern ghoul, and another as a famous slasher. People often swap tips on how to create realistic makeup or find rare fan items. The energy can be very upbeat, despite the dark themes of the films they love.

Horror Fandom Online

In online spaces, horror fans create detailed wikis about their favorite franchises, listing every character, monster power, or hidden plot twist. They post reviews on blogs or social media, sometimes filming themselves reacting to the scariest scenes. Fan fiction allows them to expand on a world's lore, imagining side stories or alternate endings.

Some horror fans run YouTube channels where they share countdown lists—"Top 10 Jump Scares," for example—or explain the backgrounds of certain myths. Podcast hosts might discuss how a film was made, the real-life legends behind it, or whether the film's villain stands for something deeper. Fans listen to these shows, learning about the genre's history and hearing opinions from people who share their interests.

Emotional Benefits of Liking Horror

Strangely enough, horror can offer emotional benefits for some people. By watching scary movies, fans can test their own reactions to fear. They might feel a sense of relief after the film ends, realizing they handled the scares. This can give a feeling of strength. It is a safe way to confront fear, knowing no real harm will come.

Also, horror can be a stress release. Some fans say that after being terrified by a monster or a tense scene, real-life worries feel smaller by comparison. They get a surge of adrenaline that burns away leftover tension from the day. Then, once the movie is over, they can laugh with friends about how they jumped at silly moments.

Horror also explores the darker side of human nature. Seeing characters make tough decisions or face moral dilemmas can help viewers think about big questions: How would I react in a crisis? Am I strong enough to protect others? Even if these situations are exaggerated, they let us think about values like courage and loyalty.

Handling Disturbing Content

Horror fans might develop ways to handle the intense material. If a scene is too graphic, they might look away or fast-forward. Some watch with the lights on or keep the volume lower to reduce sudden jumps. A few fans read plot summaries before watching, so they know what to expect. This can lessen the shock if they are sensitive to certain triggers.

Others prefer to watch with friends. That way, they can pause and chat if something becomes too disturbing. They can also check in on each other's comfort level. Horror is personal; one person might find a scene mild, while another finds it very upsetting. Good horror fans often respect each other's limits.

Horror and Younger Audiences

Horror aimed at children is usually softer. It might have spooky ghosts or friendly monsters with cartoonish looks. These shows can teach kids about bravery in small steps. As they grow older, they might move on to slightly scarier shows. Parents often decide how much horror their children can handle based on the child's temperament.

Teens often explore more intense horror. Some love the thrill of a scary story around a campfire. Others might watch a movie late at night, feeling like they are testing their courage. For them, horror can feel like a small act of rebellion or a way to share exciting stories with friends. Still, guardians need to guide them, making sure they are not watching content too extreme for their age.

Horror's Role in Pop Culture

Horror characters can become popular icons. Think of famous masked slashers or the big, silent figure who stands in the distance. They appear on T-shirts, lunchboxes, or as Halloween costumes. Even people who have never seen the films might recognize these figures from jokes or references in other media.

This mainstream visibility shows that horror is not just an underground interest. Many people who do not watch horror still know about the major villains or the big jump scare scenes that have become legendary. This can spark curiosity in new fans, who might try watching their first horror film after hearing all the talk.

Critics and Controversies

Horror sometimes faces criticism. Some argue that it is too violent or that it can harm viewers by showing disturbing acts. They worry it might inspire bad behavior. However, many studies suggest that people who watch horror understand it is fiction. They treat it as entertainment rather than a guide for real life.

Fans often defend the genre by pointing out its creative side. They see horror as a way to examine fears and human nature. They also mention that other film genres can have violence too, but horror simply places it at the forefront to provoke strong emotions. Critics might still claim certain films go too far, but the existence of age ratings helps guide viewers about what they are getting into.

The Thrill of Live Events

Some horror fans go beyond movies, attending haunted attractions where performers dress as monsters. These can be walkthrough haunted houses with dim corridors and actors who jump out at guests. Some are set in cornfields, wooded areas, or large buildings designed to look like asylums or castles. People usually buy tickets and go in small groups.

These attractions aim to scare visitors in real-time. They might use fake blood, strobe lights, loud noises, or sudden movements. The adrenaline rush is similar to a roller coaster. Fans enjoy testing themselves, seeing if they can stay cool under the tension. Once they exit, they might laugh or brag about how they did not scream (or maybe they did!).

Building Confidence and Coping Skills

Interestingly, many fans say facing fear in horror can help them cope with real anxieties. They learn to manage their racing heart or remind themselves it is just fiction. This skill can translate to real-life stress. When they feel nervous about a school test or a job interview, they might recall how they handled jump scares in a movie.

Of course, horror is not therapy, and not everyone reacts this way. Still, it can be a small part of learning how to process fear. Also, seeing characters overcome terrifying challenges can be inspiring. Even if the setting is unrealistic, the idea of finding courage in dark times can carry over to daily life.

Diversity in Horror Fandom

Horror fans come from all backgrounds. Some are older folks who grew up with classic black-and-white movies. Some are teens discovering modern ghost stories. Different cultures have their own horror legends, so people might bring unique myths and influences when they talk about what scares them. This range can lead to interesting conversations. One fan might suggest a chilling film from their country, while another recommends a local tale from theirs. Together, they broaden each other's views of what horror can be.

Negative Stereotypes

Unfortunately, horror fans are sometimes labeled as strange or overly fixated on dark topics. They might be teased or judged for their interest in gore or the macabre. But many fans lead normal lives. They simply enjoy a genre that explores fear. Some have said that the label "weird" made them hide their love for horror at school or in the workplace.

As horror has entered mainstream pop culture, these stereotypes are less strong. There is more acceptance that people can like scary movies for many reasons, and it does not make them cruel or violent. It often just means they appreciate strong emotions, storytelling, and creative visuals.

Respect Among Fans

Within horror communities, fans try to respect each other's boundaries. They might mark spoilers or warn about graphic content. They also acknowledge that some fans like lighter, spooky tales, while others watch extreme gore. By keeping a respectful tone, these communities create safe spaces for discussing fear without pressuring anyone to watch something uncomfortable.

Another aspect of respect is handling real tragedies. Sometimes, a horror story's theme might resemble a real-life event that caused pain. Fans usually agree it is wise to be sensitive when talking about such films. Horror might be pretend, but it can overlap with genuine issues like violence or mental health struggles. Thoughtful fans understand these topics need careful discussion.

Balancing Horror with Other Interests

Not all horror fans watch scary content nonstop. They might switch between a brutal zombie film and a gentle comedy next. Balancing tastes can help them avoid getting overwhelmed. Some fans watch horror primarily around certain seasons or events. Others go through phases when they crave more intense films, then switch to something else.

By mixing horror with other genres, fans keep their viewing fresh. They do not burn out on the same type of scare. This also lets them share movie nights with friends who may not like horror. In a group setting, they might pick a spooky film one week and a lighthearted movie the next. This flexibility helps maintain friendships and ensures everyone has a good time.

CHAPTER 14: FEAR IN DIFFERENT CULTURES

Fear is a universal emotion, but it does not appear the same way everywhere. Different cultures shape how people see scary events, ghosts, or unexplained forces. While "Horror Across the World" (Chapter 9) explored how countries make horror films, this chapter looks at the broader idea of how fear itself is viewed, shared, and sometimes handled in different societies. We will see that cultural backgrounds affect how people react to frightening things, what they fear most, and how they try to protect themselves from the unknown.

Cultural Roots of Fear

From ancient times, communities told stories about dangers lurking in the dark. Some cultures feared big beasts in the forest, while others worried about demons punishing those who misbehaved. These stories passed down through generations, helping people learn about local hazards or moral codes. Over time, these tales shaped how each culture approaches fear.

For example, in a land with deep forests, myths might warn of strange creatures among the trees. If the area is mountainous, residents might fear a beast that prowls the caves. Where the climate is hot and desert-like, stories might describe spirits hidden in the sands. Physical settings influence these legends, which in turn inform a group's sense of fear.

Expression of Fear

How people show fear can differ by culture. In some places, open displays of fear, like screaming or sobbing, might be seen as normal in a scary situation. People might gather, crying out together if they sense a spirit. In other cultures, fear might be kept private. Individuals try to stay calm, believing that outward panic could shame the family or break social rules.

These differences affect how horror events or stories are shared. A group that accepts loud reaction might hold bigger rituals where everyone openly shouts to ward off evil. A more reserved culture might light candles quietly at home, reciting prayers or protective words without big displays of emotion.

Taboos and Unspoken Dreads

Cultural taboos can heighten fear around certain topics. For instance, if talking about death is forbidden, ghosts might become even scarier. Or if discussing certain spirits is frowned upon, people may whisper their names instead of saying them out loud. The hush around these topics adds to the sense that they are dangerous.

In some places, people may refuse to talk about a haunting, worried it could invite real trouble. They might avoid the location rumored to be cursed, leaving it abandoned for years. The power of taboo can keep fear alive even if no one has seen evidence of the spirit. This quiet tension might be passed down through families.

Religious and Spiritual Beliefs

Beliefs about gods, ancestors, or the afterlife have a big impact on fear. A religion that views spirits of the dead as helpers might present ghosts in a kinder light. By contrast, a tradition that sees the spirit world as full of angry beings might focus on ways to appease them. Fear in these societies might center on performing the correct rituals so these forces stay satisfied.

For example, in areas where ancestor respect is strong, people might see ghosts as family members returning with advice. They are not always terrifying, though they can be if ignored. On the other hand, cultures with strict rules about demonic forces might see any unexplained event as a sign of an evil spirit. Their fear leads them to seek spiritual leaders, blessings, or protective charms.

Rituals for Protection

Many communities have rituals to ward off fearsome beings. They could involve special herbs, markings, or chants. Neighbors might gather to hold a ceremony at night, believing that unity drives away threats. Some might hang talismans on doors to keep spirits from entering. Others might burn incense or set out offerings to please wandering ghosts.

These rituals do more than chase away spirits. They also help people cope with real worries—illness, misfortune, or unknown enemies. By performing them, individuals feel they have some control. They trust that by following traditional steps, they create a safer environment for themselves and their families. This can reduce fear, even if the threat is more symbolic than real.

Superstitions and Daily Life

Cultural fear does not only arise from big events. It can appear in small traditions. Some people never step on certain thresholds, fearing it brings bad luck. Others knock on wood or avoid certain numbers. These minor acts show that fear can blend into daily habits. They might seem odd to outsiders, but they bring comfort to those raised with them.

For instance, in some East Asian regions, the number four sounds like the word for "death," so people avoid using it in phone numbers or building floors. This is not a direct horror scenario, but it reflects a deeper anxiety linked to language and tradition. If you grew up hearing "four is unlucky," you might get chills when you see it repeated somewhere important.

Family and Collective Fear

In cultures that emphasize family bonds, fear is often shared as a group. If one family member believes they saw a spirit, the others might gather around, offering support or joining in rituals to drive it away. The entire household might stay up all night, telling stories or chanting together. The fear becomes a collective experience, binding them closer.

In more individual-focused societies, a person might tackle fear alone or with a few close friends. They could watch a scary movie in solitude or talk about their nightmares privately. Without the same communal approach, their fear might feel more personal, but they also have more freedom to choose how to handle it.

Urban vs. Rural Differences

Large cities can feel crowded and fast-moving. People might fear random crime, hidden dangers in alleyways, or the isolation of an apartment where no one knows their neighbors. Rural areas, on the other hand, can have strong ties to old myths about forests, fields, or lonely roads. The quiet darkness of a countryside night can spark tales of wandering spirits.

In certain parts of the world, rapid urban growth led to new kinds of fear: fear of technology going wrong, or fear that old protective rituals no longer apply in a modern setting. Some city-based legends revolve around haunted elevators, cursed phone numbers, or strange figures seen in subway tunnels at night. These reflect the anxieties of people living in high-rise buildings or traveling underground daily.

Cultural Attitudes Toward Monsters

Different cultures have their own creatures that appear in lore. A giant serpentine being might be respected in one place, while seen as purely evil in another. Some societies have trickster spirits that punish greedy folks or misbehaving kids. Others have protective beings that scare away evil. Knowing the difference is key to understanding local fear. A monster that is purely harmful in one myth might be neutral or even helpful in another.

Sometimes, these creatures appear in children's stories, acting as moral lessons. A parent might say, "If you wander too far, the forest guardian will find you," or "If you do not finish your chores, the house spirit becomes angry." These stories keep children cautious. Over time, the kids might grow up still holding a bit of that fear. It becomes part of their cultural identity.

Festivals and Observances Related to Fear

Many cultures set aside days or periods to think about the dead, the unknown, or the boundary between life and death. People might wear outfits, light lamps, or visit graves. While some of these are solemn, others have a lively tone. Families might share meals near graves to honor ancestors or watch special performances about spirits. Though the mood can vary, the common thread is recognition that the spirit world exists or that the afterlife has a presence.

Even if these events are not meant to be purely frightening, they remind participants of the unseen. Some folks might be uneasy, especially if they are taught that on these nights, spirits draw closer. For horror fans, these times can be exciting, offering a chance to watch spooky films or share ghost tales. For others, it is simply a family tradition, a reminder of heritage and the cycle of life and death.

Fear as a Shared Language

Although cultures differ, fear can become a bridge. A scary film from one country might still frighten viewers in another, even if the local legends are unknown. Certain images—like a dark figure in a hallway—can trigger primal dread anywhere. That is why some international horror movies succeed globally. Even with cultural differences, the core feeling of dread remains.

At the same time, cultural context can add depth. A viewer who knows the local myths behind a ghost might be more scared than a foreign viewer who just sees a weird shape. Knowledge of certain taboos or superstitions can heighten the tension for local audiences. Foreign audiences might miss some layers, though they still feel the basic fear.

Tolerance for Violent Content

Approaches to horror's violence differ by place. Some societies have strict limits on gore in films, finding it disturbing or disrespectful. Others allow strong images, seeing them as a form of creative freedom. This can shape local horror. In a place with strict rules, creators might focus on psychological tension, building fear through atmosphere instead of bloody scenes. In a more permissive setting, directors may use graphic effects to shock viewers.

Audiences also react differently to violence. Some prefer to see it implied, not shown. Others want direct depiction to make the danger feel real. These tastes can connect to cultural history. A region that experienced real conflicts or tragedies might not want to see graphic violence for entertainment, feeling it is too close to painful memories.

Ghosts and Ancestors

In many Asian cultures, for instance, there is a strong link between ghosts and ancestral spirits. People believe ancestors can watch over the living, offering protection or delivering warnings. A ghost is not always evil; it might be confused or stuck. Family members may try to guide it to the afterlife with rituals. The fear is not only of the ghost itself but of failing to help a family spirit find peace.

In contrast, certain Western stories present ghosts as malevolent. If a house is haunted, it is considered dangerous, and the goal is to banish or avoid the ghost. The moral might be "leave the dead alone," or "the vengeful spirit wants revenge." Here, ghosts serve more as a threat than a relative in need of help.

Collective Trauma and Cultural Fear

Some places have faced wars, famines, or disasters that shape how they fear the unknown. A society that went through a major conflict might develop horror stories about haunted battlefields or restless spirits of those who died unfairly. The fears tap into real sorrow. Horror helps people process or talk about these events indirectly.

Similarly, a region hit by pandemics may fear diseases turning people into monsters or the concept of quarantines gone wrong. Though it is fictional, it mirrors real anxieties. These stories remind us that cultural fear is not always random. It often connects to past or ongoing hardships. By turning them into monsters or curses, folks find a symbolic way to cope.

Modern Shifts

As global media spreads, younger generations may adopt outside fear motifs. They might watch foreign horror and blend it with local myths, creating new hybrid stories. This can lead to fresh ideas but also cause older traditions to fade. Some fear that the unique local creatures will be lost if everyone only watches imported ghost or zombie tales. Others see it as a natural evolution. Culture changes over time, and fear changes with it.

People might also challenge old superstitions. Younger folks in an urban area could see older rituals as outdated, preferring modern science-based views. On the other hand, some rediscover traditions, feeling they hold spiritual truths. Either way, the tension between old and new shapes how fear is handled. This can show up in local horror films, where characters debate faith versus modern explanations.

Commercializing Fear

In some places, businesses turn fear into attractions. Spooky houses, costume parties, or fear-themed goods become ways to make money. Tourists might visit "haunted" sites for a thrill, even if locals are cautious about them. The line between real belief and staged entertainment can blur. Some older community members might disapprove, saying it disrespects the spirits or trivializes serious things.

Yet this commercialization can also keep certain traditions alive, at least in a fun way. Younger people might learn about old legends by seeing them turned into a theme park ride. While it might not be the same depth of belief as before, it keeps the stories in the public mind.

Sharing Fear Across Cultures

With social media and online discussions, people share scary tales from different cultures more easily. A user in North America can read about a Southeast Asian ghost story. A person in Africa can watch a Scandinavian horror short. This cross-cultural exchange allows people to see the variety of creatures, curses, and beliefs. They might realize that, at the core, many myths revolve around similar ideas: respect for the dead, fear of the unknown, and the danger of ignoring warnings.

This sharing can also lead to cross-overs in film or literature. A story might feature a ghost from one culture meeting a demon from another, or a family from one background moving to a land with unfamiliar lore. Such stories highlight the clash of beliefs and how universal fear can unite or alarm people, no matter their background.

Respecting Other Beliefs

When exploring fear in different cultures, respect is key. Outsiders may not believe in the same myths, but mocking them can hurt or offend locals who hold them deeply. What seems like a fun ghost tale to one group might be a serious matter to another. It is wise to approach new tales with curiosity and an open mind.

Understanding local fears can also help travelers or newcomers interact better with residents. If someone from abroad is told, "Do not go to that grove at night," it might be due to a local legend. Even if they do not personally believe it, they can show politeness by avoiding it or at least not ridiculing the concern. This shows understanding of the community's outlook.

Healthy Ways to Face Cultural Fear

Some people embrace horror or myths because it helps them face deeper anxieties. They might attend group storytelling events, where the fear is balanced by the knowledge that everyone is in this together. Or they might adapt old legends into art, short stories, or plays, exploring fear while also expressing creativity.

In recent times, counselors in some regions use local myths to talk about emotional problems. They might say, "This demon stands for your anger," or "This spirit represents guilt you carry." Such approaches blend cultural beliefs with modern therapy ideas, helping people grasp their feelings in a form they grew up understanding. Fear becomes a language for talking about personal struggles.

Ongoing Evolution of Cultural Fear

No culture stays the same forever. New threats replace old ones, and new myths arise. A community that once feared werewolves might now fear technology that spies on them. Another that once worried about farmland curses might worry more about city crime. Fear changes shape with society, always reflecting what troubles people at that time.

Still, older legends rarely vanish completely. They linger in stories told by grandparents or retold in modern media. They remain part of a cultural identity, reminding folks of where they came from and how their ancestors faced life's uncertainties. In that sense, fear is both modern and ancient, bridging the past and present in ways that keep each culture's voice alive.

Conclusion

Fear in different cultures is shaped by local beliefs, history, environment, and social norms. While everyone experiences fear, the forms it takes—ghosts, monsters, taboos, or curses—can vary widely. Some cultures openly express fear, holding group rituals or telling scary tales as a bonding activity. Others keep their dread private, using small symbols or quiet prayers to ward off danger. Religious ideas, family ties, and past traumas also play major roles in how communities handle fear.

Yet beneath these differences is a shared human trait. We all know what it is like to feel a chill when something seems off. We sense the weight of stories passed down by elders, warning us of what lurks beyond the firelight. Whether it is a vengeful spirit from old lore or a modern worry about a hidden threat, fear brings people together—sometimes in alarm, sometimes in caution, sometimes in curiosity.

By learning about how others view and manage fear, we gain insight into their values and experiences. We also discover that fear can serve multiple purposes. It can teach caution, protect tradition, spark creativity, or bring communities closer. So while the shape of fear changes across borders, its power to move us remains. It connects us to our ancestors, to each other, and to the mysteries we still do not fully understand.

CHAPTER 15: HORROR ON SMALL SCREENS

Horror is not just for theaters. It also appears on televisions and streaming platforms, sometimes in short episodes and sometimes in longer series. Many people enjoy watching scary stories in their own homes, turning off the lights and immersing themselves in tales of ghosts, monsters, or tense mysteries. Over time, the small screen has become a major outlet for horror creators, giving them a chance to spin complex plots and explore fresh angles. This chapter looks at how horror made its way onto televisions and streaming services, the history of horror shows, and the distinct features that keep audiences coming back for more.

Early Days of TV Horror

When television first became a household item, networks experimented with many genres. Some of the earliest scary programs were anthology series—shows that told a new story each episode, often with creepy twists. These drew on older radio dramas, which used sounds and imagination to frighten listeners.

Anthology Series

One early example is *Alfred Hitchcock Presents* (1955–1965), which was not always pure horror but often had dark, suspenseful stories. Viewers tuned in each week to see a self-contained plot introduced by Alfred Hitchcock himself. Another famous series was *The Twilight Zone* (1959–1964), created by Rod Serling. Although it mixed science fiction, fantasy, and moral lessons, many episodes had strong horror elements. Strange creatures or eerie events made viewers question what was real. A person might vanish into a different dimension, or a neighborhood might turn on itself under mysterious influences.

During this period, television technology was limited. Shows were in black and white, and budgets were small. But clever writers used atmosphere, dialogue, and suggestion to spark fear, rather than depending on special effects. This style matched the anthology format: you did not need recurring characters or big sets. Instead, you had a script that grabbed the audience's imagination. Even though these shows are many decades old, some episodes still scare modern viewers, proving that good storytelling can outlast fancy visuals.

Expanding Horror Themes

As TV grew in the 1960s and 1970s, horror themes spread beyond one-episode anthologies. There were shows about monsters, vampires, and the supernatural. Although not every program was strictly horror, quite a few had strong spooky elements.

Gothic Soap Operas

One notable example was *Dark Shadows* (1966–1971), a daytime soap opera. It started as a normal drama, then introduced ghosts, witches, and especially Barnabas Collins, a vampire. Audiences were hooked by the combination of daily episodes and gothic themes. Even though the special effects were simple, the show became a cult favorite. It had moody settings, old mansions, and a heavy emphasis on curses and hidden secrets. People tuned in not just for standard soap drama, but also to see if a ghost might appear or if Barnabas's secret identity would be exposed.

Made-for-TV Horror Movies

Networks also produced TV movies with horror plots. These were shorter, self-contained films shown on a specific night. Though they had modest budgets, they could still offer big scares. For instance, *Trilogy of Terror* (1975) starred Karen Black in several roles, with one segment involving a tiny, relentless doll that attacks her in an apartment. Many viewers found it terrifying, even if the doll was obviously a puppet. The close shots and frantic pacing made up for the limited resources. Such productions showed that horror on television could rattle audiences just as much as a cinema release.

The Shift to Modern Horror TV

By the 1980s and 1990s, more horror series appeared. Some took a lighthearted approach, mixing humor with spooky themes. Others aimed for serious dread. Technology allowed for better effects, and color broadcasts brought new life to monsters and gore. Networks also realized that certain audiences liked being scared weekly.

Tales from the Crypt

One standout was *Tales from the Crypt* (1989–1996), which came from a line of classic horror comics. Each episode was introduced by a puppet-like host called the Crypt Keeper, who shared puns and jokes before leading viewers into a tale of greed, revenge, or the supernatural. The show was known for mixing dark humor with vivid gore, making it distinct from earlier, tamer anthologies. It also featured well-known actors and directors, showing that TV horror could be edgy and star-packed.

Buffy the Vampire Slayer

Though it leaned toward action, *Buffy the Vampire Slayer* (1997–2003) brought horror themes to teenage audiences. It centered on Buffy, a high school student chosen to fight vampires and demons while juggling homework and friends. The show balanced scary creatures with personal drama, including heartbreak and moral dilemmas. Buffy faced monsters that mirrored real teenage problems, like fitting in or dealing with loss. Over its run, it earned a loyal following and proved that horror-fantasy could thrive on TV if it had solid characters and emotional depth.

The X-Files

Though often classed as sci-fi, *The X-Files* (1993–2002) had many episodes that were effectively horror. Agents Mulder and Scully investigated odd cases, involving creatures, ghosts, or gruesome events. A few episodes, like "Home," were so disturbing that they were controversial at the time. The show tapped into viewers' fears of the unknown—whether from space or from hidden corners of Earth. Its success opened doors for other horror-themed series, as networks saw that audiences enjoyed the thrill of weekly mysteries with a creepy edge.

Anthology Revivals

Modern television revived the anthology approach in new ways. Series like *Masters of Horror* (2005–2007) invited famous horror directors to craft hour-long stories. Each director brought a unique style, from psychological horror to splatter. The short format let them experiment with themes that might not fit a full-length movie. Audiences got to see different visions of terror every episode.

Another example is *Black Mirror* (2011–present). Though it is often described as sci-fi, many episodes rely on horror elements—like twisted technology or bleak outcomes. Viewers feel unsettled because the stories are close to real life, just one step ahead in technology. This near-future backdrop mixes modern concerns with fear, proving that horror can evolve along with society's changing anxieties.

Streaming Takes Horror Further

When streaming services became popular, horror found a perfect home. Platforms like Netflix, Amazon Prime, Hulu, and dedicated ones like Shudder gave creators room to experiment. They were not as bound by network ratings or time slots. This allowed for bolder storytelling, longer arcs, and darker themes.

Binge-Watching Horror

On streaming, entire seasons drop at once. Many viewers watch multiple episodes back-to-back. This can intensify fear because tension builds over many hours, with little break in between. If a show ends an episode on a cliffhanger—like a monster about to attack—a viewer can instantly see the next part. This format creates a more absorbing experience, pulling people deeper into the story.

Original Productions

Streaming services began making original horror series, often with high budgets and strong casts. *Stranger Things* is a popular example. It blends 1980s nostalgia, a secret government lab, and a monster-filled alternate dimension. Audiences liked its mix of friendship, family bonds, and scary creatures. Another example is *The Haunting of Hill House*, which adapted Shirley Jackson's classic novel but expanded its characters and backstory over ten episodes. Its slow-burn approach let viewers see how each family member was haunted in a personal way, building tension over time.

International Horror on Streaming

Streaming also spread horror shows from around the globe. Viewers could watch Spanish ghost stories, Korean zombie dramas, or Japanese creepfests with subtitles or dubbing. This global exchange helped fans see new styles and local myths. Shows like *Kingdom* (a Korean historical zombie series) or *Dark* (a German series mixing time travel and dread) found big audiences worldwide, proving that fear travels across language barriers.

Different Formats

On small screens, horror is not just for hour-long episodes. Some programs go shorter, with "webisodes" or short-form videos. Others do miniseries that tell a complete tale in just a few episodes. This flexibility can serve different tastes. Some watchers prefer quick bursts of fear. Others like a sprawling story that takes its time exploring characters and mysteries.

Short-Form Web Series

Certain horror stories appear on platforms like YouTube, featuring five- to ten-minute episodes. They rely on rapid scares and creative hooks to keep watchers clicking "next." This format is handy for creators with tight budgets, as they can keep sets small and focus on a single frightening concept. Fans often share these videos with friends, making them viral hits if the scare is effective enough.

Limited Series

A limited series might have around six to eight episodes, telling one cohesive story. This is a good structure for certain horror tales that would be too stretched out in a full multi-season run. It ensures the plot remains tight. Audiences know they will reach a conclusion soon, so they remain excited from start to finish. Examples range from single haunting stories to big mysteries about cursed lands or nightmarish experiments.

Character Depth in Horror Shows

One strength of horror on TV or streaming is the chance to develop characters more deeply than a film's short run time allows. With multiple episodes, the audience can see heroes struggle with personal issues, face various threats, and evolve over time. A single monster might appear across many episodes, giving viewers a chance to see every step of its effect on the community.

Long-Form Suspense

In a film, you have roughly two hours to scare viewers. On TV, you can build suspense slowly, scattering small hints across episodes. By the time the big reveal comes, the tension is very high. This method is used in shows like *American Horror Story*, where each season has a theme (a haunted house, a coven, a creepy hotel) and unravels the backstory episode by episode. The format allows for strong character arcs and unexpected twists that keep watchers coming back.

Personal Conflicts

TV horror often balances external threats—zombies, ghosts, curses—with internal conflicts among characters. People may disagree on how to handle the crisis, or they might struggle with personal secrets. In some zombie shows, the living are as dangerous to each other as the undead. This interplay of relationships and terror can deepen the horror. The audience sees that fear is not just about monsters, but also about how humans act under stress.

Changing Censorship and Ratings

Network TV in the past had strict rules about blood, gore, and disturbing material. That led to toned-down horror. Cable channels allowed a bit more freedom, but still had guidelines. Streaming services, on the other hand, can set their own boundaries, as long as they label content for mature audiences. This shift has given horror creators the chance to include stronger scares or more extreme visuals if they feel it serves the story. Some watchers appreciate the realism; others might think it can get too graphic. Either way, the small screen is no longer limited to mild content.

Horror Comedy and Mixes

Some shows blend horror with comedy, giving viewers moments of relief. Shows like *Ash vs Evil Dead* use silly gore and witty lines to balance out the fright. This style thrives on smaller screens, where fans can watch episodes casually, enjoying laughs and scares together. A comedic angle can also help engage viewers who might avoid pure horror but like an entertaining mix.

Other series mix horror with romance, detective work, or even superhero elements. For instance, *Penny Dreadful* took classic literary characters—Frankenstein's monster, Dorian Gray, and others—and set them in Victorian London, weaving horror with drama and mystery. This blend can widen the audience, drawing in those who like multiple genres.

Anthology vs. Serialized Horror

Television horror typically follows one of two structures:

1. **Anthology**: Each episode or season stands alone. Viewers can jump in at any point without needing prior episodes. This approach keeps stories fresh, as the show can shift tone or location each time. Anthologies are good for exploring many horror ideas in a single series, from ghosts in one episode to witches in another. However, building deep character arcs can be trickier, since new tales start from scratch.
2. **Serialized**: The show follows a continuous plot with recurring characters. Each episode picks up where the last left off. This approach allows for detailed world-building and character growth. The tension can build across many episodes, culminating in a big finale. The downside is that new viewers might be lost if they start in the middle. But devoted fans love the chance to see a large story unfold over time.

Both styles have loyal followings. Some watchers prefer anthologies to see new ideas each week; others love serialized arcs that keep them guessing over the long term.

Fan Interaction and Theories

Horror on small screens often sparks discussions, especially online. Viewers share theories about the latest twist or the hidden meaning of a scene. They might freeze-frame moments to spot a ghostly figure in the background. With social media, fans can talk directly to creators or cast members, giving feedback and hearing behind-the-scenes hints.

This interactive element can heighten the thrill. If an episode ends with a cliffhanger—like a main character bitten by a zombie or a demon stepping into the house—fans spend days swapping predictions. By the next episode, they gather again to see if their theories were right. This collective excitement is a big part of modern horror viewing, turning single shows into community events.

Children and Teen Horror

Television has horror content geared toward younger audiences. Cartoon channels sometimes air spooky episodes around certain times of the year, featuring ghosts or friendly monsters. These are mild and comedic, so they do not terrify children too much. Teen-focused networks also produce shows with haunted schools, vampire romances, or paranormal mysteries, balancing adventure with mild scare elements.

These younger shows can introduce kids to the thrill of horror in a gentler way. They often emphasize bravery, friendship, and learning to face fears. Although not as intense as adult horror, they help set the stage for fans who might later watch more frightening content. It is a stepping stone that blends family-friendly tales with a hint of the unknown.

International Shows and Local Myths

As streaming platforms look for unique horror stories, they turn to different regions for inspiration. A show made in Mexico might focus on a legendary creature from local folklore. A Thai series might feature ghosts from that region's beliefs. By placing these myths in a modern TV format, creators introduce them to global audiences who may never have heard of them. It also keeps local traditions alive in a new era.

For viewers, this is exciting because they discover new types of scares beyond the usual Western tropes like vampires or zombies. They might see curses attached to specific objects, ghosts that have local significance, or haunted temples that reflect real cultural architecture. The sense of authenticity can increase the fear factor, as the story seems tied to genuine beliefs.

Budgets and Quality

Some horror shows on small screens have movie-level budgets. They feature advanced effects, expensive sets, and top-tier actors. Others are done on a limited budget, relying on atmosphere and writing rather than flashy visuals. Both approaches can succeed. A huge budget can bring cinematic visuals each episode, while a small-budget show might push creativity in how it scares people.

With many streaming services competing, some are willing to spend big for attention-grabbing horror hits. They see it as a way to draw subscribers. Meanwhile, smaller production houses might cater to niche horror fans by offering unusual plots or experimental styles that bigger players might find risky. This diversity ensures that watchers can find everything from polished mainstream horror to quirky, offbeat tales.

The Ongoing Appeal

Why does horror stay strong on small screens? One reason is that people like a steady supply of fear. Instead of waiting for the next big movie, they can watch new episodes every week or binge an entire season in one go. The small screen also meets a range of tastes, from quick jump-scare anthologies to slow, thoughtful arcs that delve into deep psychology.

Another factor is comfort. Watching at home, sometimes alone at night, can heighten the fear. You sense that the horrors on screen might be just a few steps away in your own hallway. Or, if you get too scared, you can pause, turn on the lights, and calm down. This control over the viewing experience makes it easier for viewers to handle intense scenes, while still feeling the adrenaline.

The Future of Small-Screen Horror

As technology grows, new forms of horror might emerge. Some streaming platforms already experiment with interactive shows where viewers make choices that affect the story's path. Imagine picking if a character should open a locked door or run away, influencing the next scene. This could deepen the sense of involvement and fear.

Virtual reality might also come into play. Users could put on a headset and watch a horror series from inside the setting, looking around haunted rooms as the episode unfolds. This level of immersion would blur the line between viewer and participant. Though still in early stages, it shows how horror continues to adapt to fresh tools, always seeking to scare us in new ways.

In the meantime, more traditional TV or streaming horror will likely remain popular. Shows that blend strong characters, creative plots, and effective scares keep finding audiences worldwide. Some may revolve around alien horrors, while others might revolve around old curses. The variety is almost endless.

Conclusion

Horror on the small screen has grown from simple black-and-white anthologies to high-budget streaming epics. The format provides room to tell long arcs, develop characters deeply, and try different subgenres. Anthologies bring short, sharp shocks. Serialized shows build tension across many episodes. Networks and streaming services have discovered that audiences want to be frightened at home just as much as they do in theaters. In fact, the intimacy of watching in one's own space can amplify the scares.

We have also seen how streaming services support creativity. They give horror creators freedom to include bolder content and to share their work with global viewers. New technology and changing audience habits, such as binge-watching, lead to new storytelling methods that keep fear fresh. Alongside that, modern shows blend horror with comedy, romance, or other genres, pulling in fans who might not watch pure horror otherwise.

Horror on small screens is no mere afterthought compared to cinema. It is a robust avenue for innovative, chilling tales. Through limited series, anthologies, or ongoing sagas, these shows continue to shape how we experience fear. Whether you prefer short, creepy webisodes or multi-season arcs that unravel sinister secrets, there is a horror show out there ready to make you jump. As technology evolves, we can expect more boundary-pushing stories that keep the home-based audience on edge, proving once again that horror truly thrives in any medium.

CHAPTER 16: LOW-BUDGET SCARY FILMS

Not every horror movie needs millions of dollars to be frightening. In fact, some of the genre's most memorable hits were made on very small budgets. Low-budget scary films often rely on creativity and atmosphere rather than expensive sets or effects. They can leave viewers just as terrified—or even more so—than big studio releases. This chapter explores why low-budget horror can succeed, the tactics filmmakers use, and examples of titles that changed the landscape without spending much money.

Why Low-Budget Horror Works

Horror is one of the few film genres where a modest budget can actually help. Unlike giant action movies that need complex stunts or outer space sequences, horror can be effectively done with a single location and a few actors if the idea is strong.

1. Focus on Suspense
Without big funds, directors often concentrate on suspenseful storytelling. They build dread through quiet moments, eerie sounds, or shadows that hint at threats. This approach can be more haunting than flashy effects. If viewers do not see the monster clearly, their imaginations run wild, amplifying the fear.

2. Realistic Settings
Low-budget horror might film in a real house, a barn, or a run-down building, giving a grounded feel. Audiences think, "That place looks familiar," making the events more unsettling. Big-budget films sometimes use polished stages or heavy digital effects, which can create a distance. With a smaller production, you sense the rawness of the environment.

3. Creative Solutions
When money is tight, teams must innovate. They might design homemade costumes, invent clever camera tricks, or rely on minimal makeup that still looks effective on screen. This can lead to fresh ideas that stand out. For instance, if you cannot show the monster in detail, you might keep it mostly offscreen, letting viewers' fear build from the unknown.

Notable Low-Budget Classics

Some legendary horror movies were made with limited resources but still became major hits or cult favorites. They set an example that others followed, proving that scaring people does not require giant studios.

Night of the Living Dead (1968)

George A. Romero and his team made this black-and-white zombie film with a small budget. They filmed in rural Pennsylvania, using local actors and even mixing chocolate syrup with red coloring for blood. The movie's raw style made it feel uncomfortably real. Scenes of zombies surrounding a farmhouse were shocking for the time. Despite its small budget, it became a milestone in the zombie genre, inspiring countless future productions about the undead.

The Texas Chain Saw Massacre (1974)

Directed by Tobe Hooper, this film was shot on a tiny budget in the hot Texas sun. The cast and crew faced tough conditions, with the smell of real animal bones on set and hours of filming in sweaty costumes. However, that discomfort added to the movie's grim tone. The film's documentary-like style and gritty look made viewers feel they were trapped alongside the victims. Though many expected extreme gore, the movie implied more violence than it showed, relying on atmosphere to disturb audiences. Over time, it became known as one of the most unsettling horror experiences, all without lavish spending.

Halloween (1978)

John Carpenter's film about Michael Myers stalking babysitters in a small town had a modest budget. The mask worn by the killer was a simple repainted mask of a famous actor's face. Carpenter used minimal sets—mostly suburban streets and houses. The film's strength lay in its tense music (composed by Carpenter himself) and the sense that evil lurked in ordinary neighborhoods. *Halloween* went on to earn huge profits, launching a slasher wave and proving you could spawn an entire franchise with limited start-up money.

The Evil Dead (1981)

Sam Raimi and his friends financed this project partially through local investors. They filmed in a remote cabin, facing harsh weather and living conditions. The makeup and props were homemade, sometimes causing comedic mishaps, but Raimi's energetic camera work and the movie's over-the-top gore caught viewers' attention. The result was a cult sensation, leading to sequels and a TV series. Raimi's inventive style showed that passion and clever angles could overshadow budget constraints.

Found-Footage Phenomenon

One subgenre that flourished on small budgets is found footage. These films pretend to be real video recordings discovered after a horrifying incident. They are often filmed with handheld cameras, adding an authentic "home movie" vibe. This style emerged partly from financial constraints: a shaky camera and natural lighting are cheaper than big sets and stable camera rigs.

The Blair Witch Project (1999)
This is the prime example. Made for a tiny amount of money, it features three student filmmakers who vanish in the woods while investigating a local legend. The movie relies on improvised dialogue and the cast's genuine unease as they wander in the forest. Shots of the night sky, rustling trees, and strange noises in the dark create a sense of real terror. The marketing campaign also hinted that the footage might be true, sparking curiosity. The film earned massive box office returns, proving found footage could scare millions and cost very little.

Paranormal Activity (2007)
Shot mainly in a suburban house, this film focuses on a couple trying to document strange occurrences at night. The camera is often placed in the bedroom, capturing eerie moments while they sleep. Instead of major effects, the film uses quiet footsteps, flickering lights, or doors moving on their own. The tension builds as the audience fears what they might see in the next night's footage. This simple idea, executed with minimal cast and sets, became a global success. Many sequels followed, but the original's micro-budget approach remains an inspiration for budding filmmakers.

Techniques Used in Low-Budget Horror

How do these films scare audiences without major resources? Let us look at some common methods:

1. **Limited Locations**
   Sticking to one or two places cuts costs. A house, a cabin, or an abandoned hospital can serve as the primary set. Filmmakers can then explore every corner of this location, crafting tension through each

hallway or dark room. Because the viewer grows familiar with the setting, any sudden change or new clue stands out.

2. **Minimal Actors**

    Fewer characters mean less money spent on salaries. This also allows the audience to bond closely with each role. If the film focuses on a small group, each person's fear or suspicion can feel more personal.

3. **Practical Effects**

    Since digital effects can be expensive, low-budget productions often use practical methods—fake blood, homemade creatures, or creative makeup. Practical effects, if done well, can look more real than cheap computer graphics. The rawness adds to the unsettling vibe.

4. **Offscreen Threats**

    Sometimes, the scariest event is the one viewers imagine. Films might show a terrified character's reaction to something unseen, leaving it to the audience to picture the horror. This not only saves on special effects but also taps into the viewer's own mind, which can conjure more disturbing images than any budget might allow.

5. **Smart Lighting and Sound**

    Darkness and shadows can hide flaws in sets or costumes. A flickering flashlight in a dark basement can spark dread without needing an expensive monster suit. Also, sound design—footsteps, distant screams, eerie background music—triggers fear, often more effectively than big visual set pieces.

6. **Improvisation**

    Some low-budget films let actors improvise scenes. This can produce raw, realistic dialogue. It also speeds up production, as you do not need a polished script for every moment. *The Blair Witch Project* famously gave actors outlines but let them respond naturally to each other, increasing the sense of real panic.

Marketing on a Shoestring

Making the film is only part of the puzzle. How do you get viewers to watch? Big studios pay for wide advertising, but small horror filmmakers must be clever.

- **Festivals**: Film festivals are a prime way to show a low-budget horror movie. If the audience screams and then tells friends, buzz spreads. A festival award can attract distributors.
- **Online Buzz**: Social media and fan forums can be a boon. If early viewers or critics praise the scares, word travels fast. This was crucial for movies like *The Blair Witch Project*, which relied on web chatter to spark curiosity.
- **Teasers and Posters**: Low-budget creators often craft eerie posters or short teasers that hint at the scare without revealing everything. A striking image or question can hook potential watchers.
- **Grassroots Screenings**: Some filmmakers host local showings in bars, community halls, or schools. They pass out flyers, invite horror fans, and rely on word of mouth. This direct approach can help build a loyal base before the movie is sold more widely.

Modern Low-Budget Hits

Even in recent times, small productions can shock viewers and become sleeper hits.

It Follows (2014)
Though not made for pennies, it still had a modest budget compared to major Hollywood films. The concept is a curse that moves from person to person, manifesting as a slow-walking figure that only the cursed person sees. The film uses everyday suburban streets and a timeless feel, with old TVs and unclear time settings. Viewers praised its creepy atmosphere and the dread that grows from an enemy that never stops walking toward you, no matter how far you run.

Creep (2014)
Shot mostly with a handheld camera, *Creep* shows a videographer hired by an odd man who claims he has a terminal illness and wants to record a final video for his family. As they spend time together, the man's behavior becomes increasingly unsettling. The tension comes from small, strange remarks and boundary-crossing pranks that shift from awkward to menacing. With just two main actors, minimal locations, and a sense of claustrophobia, it keeps watchers on edge. The low budget does not limit the unease; it actually enhances the realistic vibe.

Advantages for Filmmakers

Directors often start their careers in low-budget horror because it allows them to learn the craft with fewer financial risks. Horror fans are known for seeking new ideas, so there is a supportive market. If the film clicks, it can attract attention from studios or bigger investors.

- **Creative Freedom**: Without a large studio controlling every choice, filmmakers can take risks, try unusual endings, or explore taboo topics. They do not have to please mainstream tastes as much.
- **Rapid Production**: Big movies can take years from script to screen. A small horror film might wrap filming in a few weeks. This fast pace means creators see results sooner and can move on to the next project if they want.
- **Cult Following**: Low-budget horror often gains cult fans who appreciate the raw style. Even if it does not make mainstream waves, it can achieve lasting fame in niche circles.

Challenges and Drawbacks

Of course, tiny budgets also bring problems. Filmmakers have to handle equipment issues, limited crew, and possibly unpaid cast members. If they need a special effect that is too costly, they might have to rewrite the script. Marketing is tough without money for major ads or billboards. Also, the final product might compete with glossy horror films that overshadow it in theaters—if it even gets a theatrical release.

Sometimes the final product looks amateurish if the team lacks skill or tries to do too much with too little. Poor sound, bad lighting, or confusing edits can hurt the scare factor. Also, if the story is not strong enough, the audience might just see the cheap side without feeling tension. So success is not guaranteed simply by having a low budget. Strong creativity and planning are crucial.

The Digital Age and DIY Horror

Modern technology helps low-budget creators more than ever. Digital cameras are affordable, editing software is widely available, and platforms like YouTube or Vimeo let people share work without pricey distribution. This environment encourages a wave of do-it-yourself (DIY) horror films.

Online Release
A film might skip cinemas altogether and launch on streaming services. If it garners positive reviews or social media chatter, it can find an audience. This approach saved the cost of physical prints or big ad campaigns. Some teams even do pay-per-view events on their own websites.

Crowdfunding
Platforms like Kickstarter or Indiegogo let fans fund a horror idea before it is made. Filmmakers pitch their concept, show a teaser, and promise perks—like a signed poster or a shout-out in the credits—to backers who donate. If enough horror lovers believe in the project, the creators gather enough money to start filming. This community-driven model fosters a sense of shared excitement, as fans feel personally invested in the outcome.

The Audience Appeal

Why do viewers seek out these low-budget horrors? Partly, it is curiosity. People wonder what fresh ideas might emerge when someone works outside the big studio system. Also, fans like the underdog story: a small team battling the odds to scare the world. Sometimes, the rawness of a cheap production can feel more genuine than a polished blockbuster. A shaky camera or a flickering flashlight can make the horror seem real rather than staged.

Additionally, low-budget horror can take risks with content. It might have an unconventional monster or a downbeat ending that big studios would fear might upset general audiences. Some fans crave these surprising twists, finding them more satisfying than typical plots.

Tips for Budding Low-Budget Horror Creators

1. **Strong Concept**
    Have a clear, striking idea that can be done with minimal resources.
    Maybe it is a single location story or a fear rooted in everyday life. The
    more unique the premise, the better it stands out.
2. **Use Natural Tension**
    If you film in a creepy house or a deserted campsite, let the setting speak
    for itself. Capture natural sounds—wind, creaking floors—and use them to
    build dread. Lean on atmosphere, not just jump scares.
3. **Focus on Characters**
    A small cast can succeed if the characters are well-developed. Viewers
    care more when they know these people's personalities and flaws. Build
    tension through interpersonal conflicts too, not just external threats.
4. **Practical Effects with Care**
    If you have gore or a creature, ensure it looks decent up close. If it does
    not, hide it with shadow or quick editing. A glimpse of a scary shape can
    be more potent than a full reveal in bright light.
5. **Audio Quality**
    Even if the visuals are rough, invest in good sound. Clear dialogue and
    effective sound effects can elevate a film. Audiences forgive grainy images
    more than they forgive muffled voices.
6. **Plan Your Marketing**
    Aim for festivals where horror is welcomed. Build a social media presence,
    share behind-the-scenes photos, or drop teasers that show your best
    scares. A strong sense of community can spark buzz.

Long-Term Impact of Low-Budget Horror

Many careers started with a small horror film. Directors like Sam Raimi, Peter
Jackson, and even James Wan (who later did *Saw*, not extremely big budget but
definitely more than pocket change) proved themselves in smaller projects
before moving to larger productions. They used the freedom of micro-budget
horror to hone their style, learn from mistakes, and show bigger investors they
could handle a film set.

Moreover, low-budget horror influences the entire genre. Big studios notice when a tiny film, made for the cost of their catering service, grosses millions. This encourages them to back new voices or adopt the found-footage style. It also pushes the boundaries of what is considered mainstream. The lines between indie and studio horror blur when a small film becomes a hit overnight.

Connection with Horror Fans

Horror fans often support low-budget projects, appreciating the effort and heart that goes into them. Conventions showcase indie movies alongside studio blockbusters. Panels discuss how small teams overcame obstacles, and fans ask questions about the homemade gore or the haunted location used in filming.

Some fans prefer the raw aesthetic, feeling it is closer to old-school horror's roots. They might collect VHS tapes of obscure slashers or cult favorites, cherishing the imperfections. The conversation around these films can be lively, with fans debating which micro-budget gem is the scariest or which cameo was best. The sense of discovery—finding a little-known title that truly frightens—adds to the thrill.

Conclusion

Low-budget scary films show that terror does not require huge funds. A clever story, unsettling sounds, inventive angles, and sincere performances can leave audiences shaking, sometimes more than big-budget productions do. By working within strict limits, filmmakers tap into creativity they might never explore otherwise. This leads to original monsters, new filming methods, and fresh ways to disturb viewers' imaginations.

From classics like *Night of the Living Dead* and *The Texas Chain Saw Massacre* to modern hits like *Paranormal Activity* and *It Follows*, low-budget horror has carved its own niche. It often turns simple ideas into unforgettable nightmares, reminding us that sometimes less is more. A shaky camera in a dark hallway or a whispered threat from beyond a locked door can ignite our deepest fears. That minimal approach, backed by strong craft, continues to win over fans who seek genuine scares rather than polished spectacle.

CHAPTER 17: YOUNG VIEWERS

Horror is often seen as an adult genre, filled with frightening images and mature topics. However, many children and teenagers show curiosity about scary films or stories. They might see older siblings watching a horror movie or hear friends at school talking about the latest creepy show. This chapter looks at how horror affects young viewers, why some are drawn to it, and the ways parents, guardians, and educators can guide children who want to explore spooky material. We will also discuss the differences between mild, kid-friendly horrors and the intense material meant strictly for adults.

1. The Natural Pull of the Unseen

Many children are fascinated by the unknown. They might tell each other ghost tales at sleepovers, scaring themselves on purpose. This interest reflects a desire to test their courage in a safe space. By hearing about monsters or haunted places, children exercise their imaginations and practice handling fear without facing real danger. Some even find it thrilling to be scared, in the same way they might enjoy a sudden drop on a roller coaster.

Because young viewers have strong imaginations, horror concepts can feel more vivid to them. A spooky figure in a hallway can stick in a child's mind for days, especially if the child already wonders about shadows at night. On the upside, a well-crafted mild scare can spark creativity, prompting children to draw pictures of ghosts or think of their own endings to a scary tale.

2. Different Levels of Scary Content

Horror aimed at younger audiences varies a lot. Some shows or books are almost playful, featuring friendly ghosts who only want companionship. Others are darker, with menacing creatures or sad backstories that can upset sensitive viewers. A big difference is how the fear is presented. Children's horror often ends on a reassuring note, letting young viewers see that the threat can be overcome. For instance, a cartoon about a haunted house might show that the ghost just wanted to be noticed, leading to a happy resolution.

In contrast, adult horror often leaves a disturbing aftertaste or includes graphic violence. Children who stumble onto these harsher tales might experience nightmares or anxiety. They can have trouble sleeping, worry about what lurks in their closet, or develop fears about the dark. Because each child is different, deciding what level of spooky content is suitable depends on both age and individual temperament.

---

3. Age Ratings and Guidance

Many countries have rating systems to help parents and guardians figure out which movies or shows are appropriate for children. These ratings consider violence, gore, language, and the presence of intense themes. Still, these systems are not perfect, because kids can differ widely. One ten-year-old might remain calm through a mild horror story, while another might be upset by the same material.

Parents and guardians can use ratings as a starting point, then watch trailers, read reviews, or ask other adults who have seen the film. Some streaming platforms also give warnings about content like violence or scary scenes. By checking these details, adults can decide if a horror show is likely to be too intense. If the child still insists on watching, the adult might view it first or watch alongside them, pausing to discuss any disturbing parts.

---

4. Why Some Children Seek Horror

Despite the potential for fright, many young viewers are drawn to horror. They might want to show they are "brave" or feel curious about the shadows and mysteries they have heard about. Sometimes, it is simply a social matter: their friends are talking about a scary movie, and they want to join the conversation. Or they might enjoy the rush of adrenaline that comes with fear, knowing deep down that the screen cannot actually harm them.

Children who enjoy scary stories can also be building problem-solving skills. In many horror tales, characters must outsmart ghosts or break curses. They might discover hidden clues or test their courage. Young viewers who see these characters succeed could feel inspired to face their own worries, even if those worries are about everyday life rather than supernatural threats.

5. Balancing Fear and Fun

For children who want to explore horror, it can be helpful to start with gentler titles. Animated specials around spooky holidays or mild ghost comedies can give them a taste of thrills without overwhelming them. If they handle these stories well, they might move to slightly scarier material, always with adult supervision. The key is to maintain a balance where the child feels excitement, not lasting dread.

Some children enjoy reading scary books with illustrations that spark the imagination. This can be safer than a live-action film, because each child visualizes the story in their own way. If a scene feels too intense, they can pause, close the book, and come back later. Also, reading often includes some emotional distance that movies or TV shows, with their loud music and sudden visuals, do not provide.

6. The Role of Parents and Guardians

Adults who care for children can play a big part in guiding them through spooky interests. First, they can talk openly about fear. A child might feel embarrassed to admit they are scared, or they might not have the words to say what disturbs them. Adults can reassure them that fear is normal and that it is okay to talk about nightmares or strange thoughts after watching a scary show.

When deciding on a film or series to watch, parents and guardians can preview it first or read detailed summaries. They can sit with the child during viewing and be ready to pause if the child gets uneasy. Afterward, they can discuss the plot, asking questions like, "What part did you find most frightening?" or "Why do you think the monster was upset?" This helps the child process the story. By explaining that it is fiction, discussing how effects are made, and focusing on the difference between screen and reality, the adult can reduce lingering fears.

7. Signs a Child Might Be Overwhelmed

Not every young viewer will handle horror smoothly. Some signs that a child
might be overwhelmed include:

- Trouble sleeping, such as nightmares or refusal to go to bed alone
- Refusal to enter dark rooms or sudden fear of places that did not bother
  them before
- Clingy behavior or jumpiness at small noises
- Changes in mood, like irritability or sadness, that seem linked to the scary
  material

If these signs appear, it might be wise to step back from horror. Reassuring the
child that they can take a break, or even skip scary content entirely, shows
respect for their mental comfort. An adult can offer alternatives, like adventure
or fantasy stories that have excitement but less fear. Over time, the child may
grow more confident and return to mild horror if they wish, but they should not
feel forced to watch scary things before they are ready.

---

8. Using Horror to Learn Lessons

Just like adults, children can find moral or emotional lessons in horror. Some
spooky tales show that kindness wins out, or that bullies face consequences.
Others highlight teamwork, with characters supporting each other against a
common threat. Children might also see heroes facing tough choices and learn
about responsibility or honesty through these plots.

In certain mild horror stories, the "monster" is a misunderstood figure who
becomes a friend when treated kindly. This can help young viewers realize that
differences or odd traits do not always mean something is evil. By pointing out
these lessons, adults can turn a simple scare into a teaching moment, helping
children see beyond the jumpy parts.

---

9. Teenagers and Stronger Content

When viewers become teenagers, their interests might shift to more intense
horror. They may watch movies rated for older audiences or talk about cult

classics. Teens often look for bigger adrenaline rushes or more complex stories
that reflect deeper social or psychological themes. They might also be drawn to
slashers or paranormal films that friends discuss.

Still, not all teens have the same tolerance. Some might be just as uneasy about
jump scares as when they were younger. Others might handle gore with no issue
but get disturbed by psychological horror. Parents and guardians can keep
communication open, asking if the teen is comfortable with the content. Instead
of outright banning certain titles (which can increase curiosity), adults might
propose watching together or setting time boundaries, so the teen does not
watch disturbing content late at night alone.

10. Horror as a Social Activity

Many teens and older children enjoy watching horror with friends. They gather
in living rooms for a movie night, sharing snacks and bracing for scares. This
group setting can reduce fear because everyone experiences the tension at the
same time. Jokes or side comments often break the tension. After the movie,
they discuss the best jump-scares or the funniest reactions.

Horror games also play a role, where friends take turns controlling a character in
a haunted house or against zombies. The interactive element can make the fear
more direct, but having buddies around helps keep it from feeling too real. Such
group experiences build memories and shared excitement, turning fear into a
bonding event.

11. The Danger of Overexposure

While some fear can be thrilling, too much can be harmful. If a child or teen
starts consuming only heavy horror, it may alter their views of the world or
deepen anxieties. They might become desensitized, no longer responding to
normal levels of fear, or they could have trouble trusting new places and people.
This does not happen to everyone, but it is a risk if frightening images become a
constant presence.

Moderation helps keep horror in perspective. A balanced media diet includes comedy, fantasy, adventure, and other genres. That variety prevents gloom or anxiety from taking over. Also, focusing on physical activities or time outdoors can balance the emotional impact of fictional horrors. Parents can gently encourage these healthy habits without shaming the child's horror interests, aiming for a constructive approach.

12. Cultural Views on Children and Horror

Some cultures hold strong taboos about children encountering scary spirits or violent material. They might see it as harmful to a child's spiritual or emotional well-being, strictly limiting any exposure. Others have traditions of telling ghost stories to children around fires or during certain seasons, passing down local myths in a family-friendly way.

Modern life, with streaming and international media, can blur these lines. Children might access horror from different cultures that do not share their family's norms. A parent might find their child watching a spooky cartoon from overseas that has traditions or creatures unfamiliar to them. This presents both risks and educational opportunities. By discussing the myths and explaining differences, adults can expand the child's worldview while still keeping a close watch on the intensity level of the content.

13. Coping Strategies for Young Viewers

If a child or teen feels uneasy after a scary show, there are ways to reduce lingering dread:

- **Conversation**: Encouraging them to talk about the frightening parts can help. Naming what scared them often makes it less overwhelming.
- **Behind-the-Scenes Details**: Showing how special effects or costumes were made can demystify the monster. If the child realizes it is makeup or computer work, they might feel calmer.
- **Comforting Environment**: Soft lighting, calming music, or favorite blankets can ease tension.

- **Creative Outlets**: Children might draw the creature or write an alternate ending, giving them a sense of control over the scary elements.
- **Daily Routines**: Sticking to normal schedules, especially around bedtime, can provide stability. If nightmares persist, gentle reassurance that the story was pretend can help them regain confidence.

---

14. Finding Age-Appropriate Material

For parents wanting to encourage a love of spooky stories without crossing a line, many kid-oriented horror options exist. Cartoon-based ghost tales, mild haunted house adventures, or short "scare" anthologies can offer excitement at a safe level. Some older family movies blend comedy with spooks, letting younger viewers laugh at silly ghosts rather than cower in fear.

Libraries often have sections dedicated to children's mystery or spooky fiction. Authors who specialize in middle-grade horror, for instance, create stories with enough suspense to thrill but not so much that it traumatizes. Checking reviews or recommendations from teachers, librarians, or trusted websites helps in picking the right choice. If an adult is unsure, they can quickly skim the book or watch an episode themselves first.

---

15. Respecting Individual Limits

Not every child or teen has the same threshold. Some find even mild ghost stories too upsetting, while others can handle more intense material at a younger age. Forcing a child to watch something they dread is unhelpful, and it may lead to long-lasting negative effects. Adults can watch for cues, like a child covering their eyes or fidgeting nervously, to see if it is time to stop.

Similarly, if a teenager says they can handle a certain level of horror, but later shows signs of distress, it is wise to have a calm discussion. A teen might not admit they are shaken, for fear of losing privileges, but they might appreciate adult support if nightmares begin. Encouraging openness and never mocking their fear is crucial. Fear is personal, and ridiculing it can worsen the emotional fallout.

16. Positive Aspects for Growth

Despite the caution needed, mild horror can have benefits for young viewers. It can teach bravery in a controlled environment, showing that fear can be faced and overcome. Children who see heroes outwit monsters might feel motivated to tackle real-life challenges. It can also spark creativity—some of the world's best writers and filmmakers started off telling scary stories in school or making small horror projects for fun.

In some cases, horror themes provide a safe space for exploring complex emotions like loss or loneliness. A haunting tale might mirror a child's feelings about missing a friend, for instance. When they see characters in a story process grief, they might gain perspective on their own troubles. This does not mean horror is therapy, but it can gently touch on issues that might otherwise remain unspoken.

17. Role of Schools and Programs

Some schools or community programs host mild haunted houses or ghost story sessions during certain times of the year. Educators typically ensure the content is suitable, focusing on fun rather than true terror. These events can be a friendly introduction to spooky themes, reinforcing the notion that not all horror is about violence or extreme shock. A school might also screen age-appropriate spooky films, then have discussions about storytelling and suspense.

Such settings provide communal experiences, so children feel less isolated in their fear. They learn that others find the same parts scary and that it is okay to jump or squeal at a sudden jolt. Teachers or group leaders can then talk about how stories are structured, from the setup of a haunting to the resolution. This knowledge can empower children to see that horror relies on certain narrative devices, which can be broken down logically.

18. Digital Age Challenges

Today's young viewers can easily find horror clips on video platforms or short scary animations. Some content might be far more intense than they expect, especially if recommended by an algorithm that does not consider age. While parental controls exist, they are not perfect, and children may still encounter shocking material.

When a child accidentally sees something too harsh, adults can step in calmly, asking what the child understood about it. A gentle talk about the difference between real-life dangers and special effects can help. If nightmares follow, addressing them quickly is best. Shaming the child for "snooping around" only pushes them away, while an open discussion builds trust. In an age where everything is a click away, guiding children through digital choices is more important than ever.

---

19. Teens Creating Their Own Horror

As children grow into teens, some want to create horror stories themselves. They might film short videos with friends, design costumes, or write scripts for school projects. This can be a productive outlet for imagination, letting them learn about plot building, cinematography, or makeup effects. Adults can guide them on managing the content if they are sharing online. For instance, they might need disclaimers or certain rating labels if the story involves violence.

This hands-on approach can transform fear into artistry. Teens see that a monster suit is just foam and paint, or that a chilling scene depends on lighting and angles. By demystifying the production process, they become more aware of how fictional horror is made. At the same time, they gain creative and technical skills that could lead to future interests in film, writing, or design.

CHAPTER 18: MODERN TRENDS

Horror has always evolved to reflect society's current anxieties and interests. Today's era is no exception. Filmmakers, writers, and fans are shaping fresh styles, mixing genres, and finding new ways to deliver scares. This chapter examines the modern trends in horror, from bold social commentary to interactive scares. We will see how technology, global media, and changing audience tastes contribute to the genre's ongoing transformation.

1. Social Themes and Real-World Issues

One notable trend is that many horror creators are tying their plots to real-world concerns. They use monsters or hauntings as symbols for larger problems. For example, a film might depict a supernatural threat that mirrors the experience of discrimination, or a malevolent entity that represents environmental harm. This approach is sometimes called "social horror," because it shines a light on society's troubles.

- **Race and Identity**: Recent films have explored how prejudice and fear of others can produce real terror. Stories about a haunted house may also tackle the history of oppression in a certain neighborhood, linking ghosts to past injustices.
- **Economic Struggle**: Some horror tales focus on families dealing with poverty or unfair housing, showing how desperation can lead them into dire bargains or cursed properties.
- **Climate Anxiety**: With the growing worry about extreme weather and environmental damage, horror has begun to present floods, hurricanes, or mutated creatures as reflections of nature's fury.

By weaving social issues into frightening plots, modern horror gives viewers both scares and a chance to reflect on deeper themes. This blend can make a film resonate long after the final scene.

2. "Elevated Horror" and Complex Stories

In recent years, critics have used the term "elevated horror" to describe films that focus on slow-building tension and emotional drama, rather than standard jump-scares. These movies might dive into grief, family conflict, or hidden trauma, using horror elements to intensify the drama. The threat may not be a simple monster but a deeper psychological dread.

- **Hereditary**: This film examines family secrets and grief, mixing in supernatural forces to highlight how past wounds can destroy a household from within.
- **The Witch**: Set in colonial times, it explores isolation and religious paranoia. The slow pace and bleak setting build a sense of dread, with the devil lurking in the woods.

While some fans love these thoughtful approaches, others miss the lighter or more direct scares of older horror. Still, the success of these layered stories proves the genre can tackle mature ideas while remaining terrifying.

---

3. The Rise of Short-Form Horror Online

Online platforms support short-form videos that deliver quick bursts of fear. A single creepy clip, two minutes long, can go viral and reach millions. These shorts often feature a simple premise: a shadow behind someone, a strange figure in a mirror, or an unexplained movement at the edge of the screen. The short format means they do not need a deep backstory, just a punchy scare.

Independent creators can produce these videos with minimal gear. Social media allows them to spread fast, and fans share them with friends. Some creators build followings by releasing a new short scare every week, hooking viewers who crave a tiny jolt of adrenaline. This trend shows that horror can thrive outside full-length films or TV episodes. The essence of fear can be captured in brief moments if done cleverly.

---

4. Found Footage Keeps Evolving

Though it started decades ago, the found-footage style remains popular, with modern twists. New gadgets, from smartphone apps to live-streaming drones, give creators fresh ways to simulate "real" recordings. A horror story might unfold as an internet influencer visits a haunted location and streams the experience, letting viewers at home watch events in real time—until something goes terribly wrong.

Some films mix found footage with regular cinematography, creating a layered narrative that jumps between a polished viewpoint and shaky first-person shots. The key remains the same: blur the line between fiction and reality. Audiences feel as if they stumbled upon actual events, which can heighten the tension. However, found footage can also face criticism if it does not feel authentic or if viewers tire of shaky camera angles.

5. Streaming Dominance

Modern horror often finds its biggest audiences through streaming services rather than theaters. Many fans prefer to watch scary stories at home, on their own schedule, sometimes bingeing entire seasons in one or two nights. This shift changes how creators design their stories. A television series might insert cliffhangers at the end of each episode, encouraging viewers to keep going. A film might release online only, skipping cinemas entirely.

Streaming platforms also seek out exclusive horror content to attract subscribers. They might fund original productions by emerging directors or buy the rights to foreign horror films, bringing them to international attention. The result is a continuous flow of new horror titles, often popping up with little warning. Horror fans who frequent these platforms can discover gems from around the world, widening their perspectives on what is scary.

6. Interactive and Immersive Experiences

Some horror projects involve the viewer directly. These can be interactive films where you choose the characters' actions, changing the plot's direction. Others might be virtual reality (VR) experiences, placing you inside haunted houses or letting you face monsters as if they were real. A jump scare in VR can be more intense than on a flat screen, because you feel surrounded by the environment.

Beyond digital technology, immersive theater events place participants in horror settings. You might explore a creepy building, guided by clues, while actors in costume pop out at unexpected moments. This live approach merges elements of stage performance and haunted house attractions, giving a unique sense of presence. As technology improves, we can expect even more immersive horror events, where the line between audience and performer continues to blur.

---

7. Horror Comedy Grows

Mixing humor and terror has been around for a while, but modern horror comedy has found fresh audiences. Movies and series that pair shocking scenes with jokes or absurd plots appeal to those who want both laughter and adrenaline. This approach can help viewers handle fear. After a scare, a funny line or silly moment lets them breathe, preventing emotional overload.

Recent examples include films where zombies become the backdrop for slapstick shenanigans or where vampires live mundane lives in modern apartments with comedic misunderstandings. Social media encourages the spread of funny horror clips too, as people share unusual moments. Horror comedy often invites a broader crowd, including those who avoid purely frightening material but enjoy dark humor. This blend shows that horror can be flexible and not always grim.

---

8. Global Exchange of Horror Stories

Thanks to the internet and streaming, viewers can explore scary tales from many cultures. This global exchange leads to creative crossovers. A director in one country might borrow a spirit legend from another region, blending it with local myths. Or a streaming service might commission a series based on a best-selling foreign horror novel, adapting it for worldwide audiences. Fans watch new stories that expand their horizons beyond the usual ghosts or slashers.

Some results include:

- **Korean Zombies**: Shows like *Kingdom* introduced historical settings and unique themes.
- **Latin American Folklore**: Movies about cursed jungles or shape-shifting beings from regional myths.
- **African Supernatural**: Tales about witches or haunted villages shaped by local traditions.

When these stories reach wide audiences, they influence each other. A horror writer in Europe might be inspired by an Asian ghost film, or vice versa. This cross-pollination ensures that horror stays fresh and surprising.

9. Larger Budgets and Franchise Building

While low-budget horror remains strong, there is also a trend of major studios giving big budgets to new horror films. These productions feature famous actors, advanced special effects, and expansive marketing. Some aim to start franchises with sequels, prequels, and spin-offs. A single monster or haunted universe can branch into multiple stories, each earning money for the studio.

Some fans appreciate the higher production value, enjoying polished visuals or epic scale. Others worry that bigger budgets might water down the creativity that is often found in smaller, risk-taking horror. Still, when done well, a large-scale horror film can deliver memorable chills and attract a wide audience. The success of such movies shows that horror can be profitable, not just a niche interest.

10. Remakes and Nostalgia

A modern trend is remaking or rebooting classic horror. Filmmakers revisit old stories, using updated effects or different angles on the plot. Fans of the original might be curious to see how the new version compares. Nostalgia plays a big role, as people recall how they felt watching the older film. They look for that feeling again, sometimes with a modern twist.

- **Changes in Tone**: Some remakes aim for a darker, more realistic style, removing campy elements from the original.
- **Gender or Story Swaps**: Others shift the perspective to a new main character or alter key details, keeping the core idea but modernizing the themes.
- **Mixed Reactions**: Critics sometimes say these remakes do not capture the spirit of the first version or are unnecessary cash-ins. Yet many viewers still show up out of curiosity, making these films successful enough to keep the trend alive.

---

11. Technology as a Threat

Horror often mirrors real fears, and technology is now a big source of anxiety. Modern tales show apps that summon ghosts, AI that becomes hostile, or VR games that trap users. These stories question our reliance on gadgets. They ask if we risk losing control when machines and data control so much of daily life. A simple phone glitch can become the start of a terrifying chain reaction in these plots.

Movies might focus on social media horrors—like a viral challenge that turns deadly or an online presence that stalks victims. Others revolve around hacking or smart homes gone rogue, where the locks and cameras are no longer under human command. The theme resonates with modern audiences who see constant updates, privacy breaches, and the fast pace of tech.

---

12. Shifting Taboos and Freedoms

Society's view of what is too shocking changes over time. Material once considered too extreme might now be allowed, depending on the region and rating systems. This can lead to more explicit gore or psychologically disturbing content in mainstream horror. Some creators use that freedom to explore heavy subjects like abuse or mental illness in a raw manner.

However, not all modern horror aims for brutality. Some stories choose subtlety, focusing on dread instead of graphic imagery. The point is that creators have wider choices than before. They can craft a quiet ghostly atmosphere or show explicit violence, depending on what best serves their story. Audiences can pick and choose from a broad range of styles, from gentle spookiness to hard-edged terror.

13. Ongoing Interest in Vampires, Zombies, and Classics

Certain classic creatures—vampires, zombies, and werewolves—never truly leave the horror scene. They evolve with each new generation. Modern vampire tales might depict them as social media influencers or bored immortals living in big cities. Zombie stories mix survival elements with personal drama, sometimes adding a scientific spin on infection. These old monsters stay relevant by shifting them into modern contexts, addressing issues like morality, loneliness, or group panic.

Even ghosts get updates. A haunted house might now be a haunted website, or the spirit could manifest through an online chat session. Writers and directors keep reimagining these timeless concepts, finding fresh angles to captivate viewers who think they have seen it all.

14. Empowered Characters

Another shift is the portrayal of more capable and proactive heroes. Rather than the old trope of helpless victims, modern horror often features protagonists who fight back against evil. They might be skilled survivors, clever researchers, or individuals driven by a personal mission. Some creators also strive for better diversity in casting, showing different backgrounds and experiences.

In older films, certain groups—like women or minorities—were reduced to stereotypes or quickly disposed of. Now, we see stronger and more nuanced roles. The "final girl" concept, once a cliche in slashers, evolves into a complex lead who chooses her destiny rather than stumbling into it. Empowered characters do not remove fear; they simply increase the tension by showing that even strong, prepared people can be challenged by overwhelming threats.

15. Collector's Culture and Extended Universes

Horror fandom is bigger than just watching films. Modern fans buy collectibles, from detailed figurines of famous monsters to limited-edition posters. They follow interviews with directors and study behind-the-scenes features. Some series develop extended universes with prequel comics, spin-off novels, or digital shorts that fill in story gaps. Fans who love deep lore can spend hours analyzing each detail.

This engagement can raise hype for new releases. When a film or show teases a small connection to a larger universe, fans speculate online, feeding the excitement. Studios recognize this devotion and may hide "Easter eggs" referencing other projects. It is a cycle: devoted fans keep the buzz alive, and creators reward them with more layers to explore.

---

16. Mental Health and Trauma-Focused Horror

Horror increasingly looks at mental health issues, portraying how depression, anxiety, or guilt can shape experiences. A character's ghost sightings might tie to personal grief rather than an external spirit. Or a haunted location could represent a suppressed memory. This approach speaks to viewers who see reflections of their own emotional battles in the characters' struggles.

Such narratives can be intense, pushing horror into a realm where the biggest scare might be inside a person's mind. The monster becomes a metaphor for feeling trapped, worthless, or overwhelmed. While this can be cathartic for some, it may be triggering for others who find the subject too close to home. Nonetheless, it shows the genre's maturity, as it tackles topics once avoided or dismissed.

---

17. Smaller Screen, Bigger Impact

We have talked about streaming, but it is worth noting how mobile devices also reshape how people watch horror. Some fans consume short videos on their phones, even in bright daylight. Others stream entire films on tablets with headphones, intensifying jump scares in public places. Horror creators now consider these viewing habits, possibly structuring stories that fit smaller screens or short attention spans.

At the same time, advanced home theaters let viewers see high-definition images on large TVs with surround sound. This can rival the cinema experience, especially for jump scares and moody ambience. So, horror can be watched in many ways—on a phone in a busy cafe or a big screen in a silent living room. Each environment changes how viewers process fear.

---

18. Collaborations with Other Genres

Modern horror does not stand alone. Directors and writers blend it with romance, comedy, historical drama, or sci-fi. A love story might happen in a town plagued by werewolves. A detective show could feature occult murders that slowly become a full-blown demonic invasion. These cross-genre projects appeal to fans who like layered storytelling. They keep horror from becoming repetitive.

A science-fiction film with horror elements might explore cosmic dread or monstrous alien life, similar to classic "creature in space" plots. Meanwhile, a period drama set in the 1800s might add ghosts haunting a stately mansion, weaving historical details with spooky corners. Blending genres can also broaden an audience, luring people who typically avoid pure horror but enjoy the mixed style.

---

19. Festivals and Fan Events

Horror festivals have gained momentum worldwide, not just for movies but also for fan culture. People attend panels, meet creators, and watch exclusive screenings. This fosters a sense of community—viewers celebrate their shared passion and keep an eye on emerging trends. Conventions also highlight new indie films or experimental shorts, giving them a boost in recognition.

Some festivals focus on particular themes, like extreme gore or psychological terror, drawing fans who crave those subgenres. These events provide a testing ground for new talents. A short film that impresses festival-goers might get picked up for expanded development. As a result, the festival circuit shapes which projects become the "next big thing" in horror, encouraging creators to experiment.

---

20. Conclusion

Modern horror is alive with innovation and variety. From thoughtful "elevated" stories that delve into human emotions, to short viral clips that shock viewers in seconds, the genre spans countless approaches. Technology enables interactive experiences and immersive shows, while streaming services welcome fresh voices and global tales. Social commentary, mental health reflections, and new spins on classic monsters ensure that horror never stagnates.

Audiences now expect more than a basic jump scare. They want meaningful themes, deeper characters, or at least a clever premise. This push for quality does not erase the desire for pure entertainment—comedic twists, thrilling chases, and nostalgic remakes still hold strong. Instead, it expands horror's reach. It can address real-life worries, amuse with dark humor, or transport us to other realms filled with nightmares and wonders.

As we advance further into the 21st century, we can expect horror to continue merging with future technology and global influences. Each new generation will bring fresh anxieties to the screen, turning them into chilling stories we share in theaters, living rooms, or on devices we carry everywhere. Modern horror does more than scare: it reflects who we are now—our hopes, our fears, and our endless fascination with the things that go bump in the night.

CHAPTER 19: ENDURING POPULARITY

Horror has lasted through centuries, appealing to audiences across time and different societies. While other trends can fade, scary stories keep reappearing in new forms. From ancient folk tales told around fires to modern digital content, the urge to share frightening ideas is strong. This chapter looks at why horror remains so popular, the factors that keep it alive, and how it affects both casual viewers and dedicated fans. We will see that horror's enduring attraction is not just about monsters or violence, but also about deeper human needs like facing the unknown and connecting with others through strong emotions.

1. Timeless Themes

One reason for horror's steady presence is that its core themes never grow old. People always wonder about death, the afterlife, or mysterious forces. Since these questions do not have simple answers, scary stories speak to deeply held concerns that remain over generations. A monster or ghost can symbolize human fears of mortality or guilt. Even if the creature changes shape—an ancient spirit in one era, a mutated beast in another—the basic question stays the same: what if something beyond our control lurks nearby?

These timeless worries unite people. A person living in a city with modern technology can still relate to a legend about an isolated cabin in the woods, because the feeling of being threatened is universal. Horror storytellers tap into that collective sense of unease, showing that no matter how much progress we make, we cannot escape fundamental questions about life and danger. This gives horror a long life, as each new generation finds a way to show old fears in a new light.

2. Emotional Intensity and the Need for Thrills

Humans often seek strong emotional experiences. Many are drawn to sports for excitement or thrill rides for that rush of adrenaline. Horror taps the same desire, offering a rush of fear in a controlled way. When we watch a scary movie or read a frightening story, we can press pause or put down the book. Our heart rate might spike, but deep down we know we are safe. This sharp feeling can be addictive, encouraging people to return for more.

Additionally, horror can break the routine of daily life. The jolt of tension, the sweaty palms, the shortness of breath—these stand out from normal day-to-day experiences. Afterward, viewers feel a sense of release or relief when the danger on screen is over. For some, it becomes a bonding experience to watch with friends, all jumping at the same scene and then talking about it afterward. This combination of fear, safety, and shared excitement helps keep horror fresh and in demand.

3. Cultural Reinvention

While horror deals with timeless fears, it also shifts with social values and trends. In different periods, certain horror styles gain popularity. For instance, in times of global uncertainty, viewers might watch more apocalyptic tales. After major social changes, new subgenres might rise that comment on the era's tensions. This flexibility helps horror remain relevant. By adapting old ideas and bringing them into modern settings, creators allow the genre to speak to current issues.

Such reinvention takes many forms. A classic vampire story might be updated to focus on social isolation in a digital age, or a ghost tale might address climate problems by linking a haunting to an ecological disaster. Even monster designs change over time, reflecting new technology or fresh cultural symbols. Because horror can quickly respond to what worries people now, it stays meaningful instead of becoming stuck in the past.

4. Shared Rituals and Social Bonds

Horror is often enjoyed in groups, whether in a theater, at home with friends, or online with chat discussions. This group setting heightens the effect. Everyone gasps together at a jump scare or debates the film's hidden meanings. This creates a sense of community that can last beyond the credits. People discuss theories, debate the scariest moments, and relate the story to personal experiences.

In some traditions, family members might gather around to tell ghost stories on specific nights. In modern contexts, fans hold watch parties, dressing up and making special snacks. This sense of shared ritual ties horror to cultural traditions that bring people closer. Even children's mild ghost tales can build a sense of belonging. Over time, these events become memories that pass on the love of scary stories from one generation to the next.

5. Technology Expands Reach

In earlier times, fear-based tales spread by word of mouth or written text. But modern technology has magnified horror's impact. Movies, TV shows, streaming, and social media help stories travel fast. A chilling indie film can gain a worldwide audience within days if it gets good reviews or goes viral. Clips of terrifying scenes are shared on platforms, sparking curiosity.

Also, advanced technology allows for immersive experiences. Virtual reality, interactive games, and phone-based horror apps let fans go deeper than ever before. These innovations draw in new viewers who might have avoided the genre in simpler formats. By continually embracing the next wave of tech—be it streaming or augmented reality—horror creators keep the content feeling fresh and accessible.

6. Freedom to Experiment

Horror is one of the few genres where risk-taking is often rewarded. Viewers can be open to new storytelling methods, unusual visuals, or strange plot twists, as long as the result feels scary. This encourages creativity. Filmmakers who might struggle to pitch a mainstream drama can find success with a bold horror concept. Even a small studio can make a big impact if their idea is original and executed well.

Because of this creative space, we see all sorts of subgenres thriving: psychological horror, sci-fi horror, found footage, or comedic horror. Writers and directors can layer in comedic elements, deep social commentary, or surreal visuals. Fans may seek out these experiments specifically, enjoying the variety.

With so many different ways to present fear, the genre keeps renewing itself, guaranteeing it does not become dull or predictable.

7. Catharsis and Coping with Real Worries

Horror provides a means for audiences to confront harsh realities indirectly. Some folks watch a terrifying scenario and see parallels to their own anxieties—about health, personal loss, or broader social issues. By surviving the story's tension, they can experience a form of catharsis, releasing pent-up fear. This does not solve real problems, but it can help people feel less alone. They see that fear is a shared human emotion, and facing it onscreen can be empowering.

Writers often use symbolic monsters to represent intangible threats. A shapeless entity could stand for a global concern, like an invisible illness or a creeping social divide. In seeing characters fight or escape that entity, viewers imagine themselves as capable of handling their own problems. Even if the story is fictional, the emotional response can be meaningful. This coping aspect, subtle though it is, adds to horror's enduring draw.

8. Academic and Critical Interest

Horror has also gained acceptance in academic circles. Scholars examine older horror movies for historical context, discussing how they reflect the fears of certain times. They may analyze how monsters link to social prejudices or how ghost stories change when cultures shift. University courses on film or literature often include horror units, introducing students to the complexities behind the screams.

This academic acceptance elevates the genre's status. Horror is not just "low entertainment" or exploitative material. It can be studied as an art form that challenges viewers' comfort zones and explores philosophical ideas. That in turn encourages more serious creators to enter the field, producing horror that has both emotional punch and intellectual depth. With such scholarly backing, horror gains legitimacy, bolstering its long-term presence.

9. Mainstream Blockbusters and Revenue

Another aspect sustaining horror is its financial success. Horror movies often require smaller budgets than action or fantasy films. Yet they can perform extremely well at the box office if they strike the right tone. A well-marketed horror flick can earn back its costs multiple times over. Studios notice this profit potential, so they continue to invest in new scary projects.

Even when the global economy struggles, audiences still seek out horror as a form of escape or emotional release. The genre's relatively low production costs also let smaller companies compete, leading to a steady flow of new titles. Some may be quickly forgotten, while others become cult hits or mainstream sensations. Either way, the consistent revenue stream ensures that horror remains a safe bet in the entertainment world.

---

10. The Influence of Fandom

Horror fandom contributes significantly to its continued popularity. Devoted fans attend conventions, write blog posts, and create online communities where they dissect every detail of their favorite titles. They support indie projects through crowdfunding, champion lesser-known films, and keep track of upcoming releases. This loyalty keeps the genre vibrant and fosters a sense of belonging among fans.

Some fans collect merchandise, from posters and statues to limited-edition DVDs. Others produce fan art or cosplay as beloved characters. The communal aspect means that interest does not fade once the movie ends; it evolves into fan theories, spinoff stories, and continuous engagement. This grassroots excitement can boost a film's profile, turning obscure works into iconic classics over time.

---

11. Cycles of Reinvention

Horror has gone through many cycles. For a while, supernatural horrors might dominate, then slashers might take over, followed by zombie crazes or paranormal thrillers. Each wave can fade when viewers grow tired of certain tropes, but soon a new wave appears, perhaps revisiting an old style with a twist. These cycles show the genre's adaptability. It learns from past phases, merges them with new ideas, and returns stronger than before.

For instance, the 1980s were known for slasher icons, the 1990s brought meta-awareness in teen horrors, and the 2000s saw found footage become a major trend. More recently, we have seen psychological and "elevated" horrors focusing on deeper themes. Through each shift, the genre remains active. Even older methods, like classic gothic sets or black-and-white style, can resurface if done in a fresh manner.

12. Personal Identification with Characters

Another factor is the empathy viewers feel for horror characters. A well-made horror piece often features ordinary people in extraordinary danger. Audiences can picture themselves in that situation. They root for the characters to survive and share in their relief or despair. This identification deepens the experience, making each close call or horrifying discovery feel real.

Some modern horrors go further by building complex leads with personal struggles. Instead of a simple hero, we have someone dealing with mental health, guilt, or broken relationships. Watching them battle external and internal threats together can be emotionally gripping. Horror with strong characterization stands out, leading fans to revisit it over time. This rewatch factor supports the genre's staying power, as viewers form connections that last beyond a single screening.

13. Nostalgia for Past Classics

Old horror movies and shows often age into "classics" that new fans discover and adore. The sense of nostalgia—either from personal childhood memories or from the cultural legacy these works hold—draws people. Festive times might inspire re-watching famous slashers or vintage monster films. This cycle of rediscovering past titles feeds the genre, ensuring that the older library does not vanish.

Studios also produce remakes or sequels to older horrors, sparking renewed interest in the original. People watch the remake to compare, or the sequel to see how the story continues. Even if the new version divides opinions, the conversation stirs excitement around the brand. Thus, nostalgia acts like a steady current that keeps older horror relevant in modern times.

14. Accessibility Through Multiple Mediums

Horror is not just about movies. It thrives in books, podcasts, graphic novels, video games, and live events. This variety of platforms multiplies its reach. A gamer might experience interactive horror in a storyline that demands personal decisions. A reader might get chills from a slow-burn supernatural novel. A group might attend a haunted attraction during a holiday season. Each medium adds new angles.

Because fans can switch between these forms, they rarely run out of fresh scares. If they tire of jump scares in movies, they might pick up a horror board game or listen to chilling audio dramas. This cross-pollination broadens the audience. Some people who rarely watch horror films might still enjoy thriller novels or comedic horror-themed TV episodes. Hence, the genre's popularity grows through multiple channels.

15. The Power of Word of Mouth

Horror often benefits from word of mouth. Friends whisper about a shocking twist or a terrifying scene, urging others to see it. Curiosity can be a strong motivator. People want to know if they can handle the fear that their peers mention. This kind of informal marketing can be more persuasive than commercials, as it comes from someone the potential viewer trusts.

In the digital age, word of mouth expands through social media, forums, and reviews. A single frightening trailer might go viral, with users reacting in real time. Enthusiasts share links or highlight timestamps of the scariest moments, and onlookers check it out just to see if it lives up to the hype. When a small horror film impresses early watchers, it can explode in popularity almost overnight. This phenomenon shows that genuine reactions, not just big ads, fuel horror's presence.

16. Psychological Comfort in Facing Fear

Facing fear in a safe space can be oddly comforting. When life outside feels uncertain, a horror story provides a structured set of fears. It has a beginning, middle, and end. Often, the characters confront or flee the threat, achieving some resolution. Even bleak endings at least contain the terror within the story. By seeing fear contained in a narrative, some viewers gain a sense of mastery over chaos.

This process can be cathartic. People watch characters endure horrifying events and then reflect on their own problems, realizing those problems might not be so bad. Or they feel empathy for characters who face unknown evils. In a way, horror stories let us mentally rehearse how we might deal with life's darker aspects. This sense of preparation, even if partial or symbolic, draws many back to the genre.

17. Diverse Voices and Representation

Modern horror is becoming more inclusive. Creators from various backgrounds bring different cultural myths, new angles on old monsters, or perspectives shaped by personal experiences. This diversity makes the genre richer, attracting broader audiences who see their own stories or heritage reflected. A tale set in a lesser-known location or featuring mythologies outside the standard Western canon can spark fresh interest.

Likewise, characters once sidelined or reduced to stereotypes are now taking center stage. When viewers see leads who look like them or share their background, they feel a stronger connection. This inclusive push aligns with broader social trends, but it also benefits horror. New voices create stories we have not seen before, keeping the genre from becoming stale or repetitive.

18. Continuous Discussion and Debate

Part of the fun in horror's endurance is the endless debate it sparks. Which movie is the scariest? Which monster is the most terrifying? Are jump scares cheap or effective? Fans can discuss for hours, referencing examples that prove their points. They might argue about deeper meanings in psychological horror or defend a guilty-pleasure slasher that others mock.

Such debates do not have definitive answers, so they remain lively. Each new release adds more material to the conversation. Over time, the genre builds a large pool of references, from classic black-and-white films to modern streaming hits. Viewers can rank them, analyze them, or connect them to broader contexts. This ongoing chatter keeps horror alive in the public mind, ensuring it never slips quietly away.

19. Evolving Monster Archetypes

Horror thrives on memorable villains and creatures—vampires, werewolves, ghosts, zombies, demons, and more. These archetypes adapt to fit the era. Today's vampires might struggle with technology or moral questions about feeding on humans. Zombies may appear as victims of lab-made viruses, reflecting concerns about science or global crises. Ghosts could be linked to digital curses or haunted social media pages.

By evolving, these archetypes remain relevant. Newer creations also emerge, like twisted artificial intelligences that threaten people, or supernatural reflections of cyberbullying. The monster's shape might change drastically, but the fear underlying it stays. Each time an archetype is reinvented, fans get a fresh reason to revisit or discover it. This cyclical process ensures that horror creatures keep capturing imaginations.

CHAPTER 20: FINAL THOUGHTS

We have traveled through the many facets of horror—its roots, psychology, cultural ties, evolving subgenres, and the people who enjoy it. Now it is time to wrap up these ideas, bringing them together to provide a broader understanding of why we are captivated by the scary side of fiction. In these final pages, we will reflect on the main points covered, revisiting how horror touches our fears, tests our emotional limits, and unites us around the thrill of being safely terrified. We will also consider the future of horror and how it may continue to change along with social trends and technological advancements.

1. Horror as a Window to Human Nature

One of the biggest takeaways is that horror is more than shock or gore. At its best, it serves as a window to human nature. It exposes our vulnerabilities, revealing how we react when we meet the unknown or cannot rely on normal rules. Monsters and ghosts are not just outside threats; they can stand for inner struggles like guilt, regret, or suppressed anger. A cursed object might reflect the harm caused by greed. A haunting could show how family secrets weigh on us.

Seeing characters handle these challenges lets us ponder our own responses. Would we run, fight, or break down if faced with a similar horror? That question, though hypothetical, resonates in real life. We might not meet a vampire, but we do face moral choices, hidden fears, or situations that feel out of our control. Horror dramatizes these moments, giving us a form of rehearsal for emotional trials.

2. The Thrill of Fear

Horror also explains why some people crave strong emotional rides. Fear, when managed, can be exciting. Our bodies react with adrenaline, quickening the heart. We experience shock, then relief as the danger ends. This cycle of tension and release is enjoyable for many. Horror stories let us practice fear safely, reminding us that we can endure stress and come out okay.

This thrill can be shared. When friends watch a ghost film at midnight, they jump together at every scare. Then they laugh, talk about it, or mock the monster's design. The closeness that emerges from facing fear together is special. As a group, we feel braver. This social dimension has been there since ancient times, when people told scary tales around fires, drawing closer as they faced the possibility of lurking beasts outside.

3. Endless Styles and Subgenres

The horror field offers countless subgenres: slashers, supernatural tales, cosmic dread, body horror, psychological suspense, and more. Each style caters to distinct tastes. Some want a gory spectacle, while others prefer subtle tension. This variety keeps horror fresh. If viewers tire of jump scares, they can shift to moody ghost stories. If they want comedy with fright, they can find horror comedies that blend laughs and shocks.

Across this variety, creators continue to experiment, shaping new hybrids or reviving old approaches. A fresh wave might revolve around technology gone wrong, or take place in historical settings, or mix with romance or detective plots. The result is that horror never stands still. Every generation sees new angles that keep fans curious. Those who love the genre know they will never run out of unsettling plots.

4. Reflecting Social Changes

Throughout this book, we noted that horror echoes real anxieties. When people worry about medicine, we see zombie outbreaks or dangerous labs. When technology becomes central, we get haunted apps or sinister AI. When social divides widen, we might see plots where neighbors turn on each other under supernatural pressure. Horror intensifies what is already bothering a community, giving it a tangible shape—a demon, a curse, or an evil figure.

By exploring these fears in a fictional space, we can face them indirectly. Some stories even point to possible solutions or moral lessons, although not all do so overtly. Viewers may see themselves in the characters who confront social issues, feeling validated that their concerns matter. This connection to real life can turn a scary film into a broader commentary, letting horror be both entertaining and thought-provoking.

5. The Role of Community

Horror's popularity is linked to how it forms communities. Conventions bring together fans who share interests in slashers, creature features, or spooky novels. Online groups discuss theories, rank the scariest scenes, and trade recommendations from around the world. This group aspect can be welcoming. Everyone is there to be frightened and to enjoy it, removing barriers that might exist in everyday settings.

For many, horror communities become a place of belonging. People can celebrate creativity through costumes, artwork, or fan fiction without judgment. They also bond over the memory of that first time they watched a classic film or overcame a fear by reading a certain novel. These interactions reinforce the notion that horror is not just about dread—it is about unity and expression.

6. Guidance and Caution for Younger Audiences

We talked about how younger viewers might be drawn to horror, either from curiosity or peer influence. It is important to note that while mild scary stories can help children explore courage and imagination, pushing them toward overly intense material can cause anxiety or nightmares. Parents and guardians who guide children well allow them to enjoy the excitement without prolonged distress. If a child sees that they can handle a small scare and still feel safe, they gain confidence.

However, not all children want or need horror. Some do better avoiding it. The key is open communication: if a young person is intrigued, adults can pick age-appropriate choices, watch together, and discuss any fears. This approach ensures that horror remains a positive influence rather than a source of trauma. Teens can eventually branch out into deeper horror, but they should be free to set their own comfort levels.

7. The Growing Influence of Technology

We live in an era where technology molds every aspect of life, and horror is no exception. Digital tools allow for convincing special effects at lower costs, letting even smaller productions achieve strong visuals. Streaming platforms bypass traditional gatekeepers, giving independent creators a chance to shine. Social media helps small movies explode in popularity if early viewers rave about them.

Additionally, interactive horror—through VR, AR, or branching narrative games—presents new ways to scare audiences. People can move within a virtual haunted house, or pick story paths that influence who survives. Such immersion can heighten fear, erasing the barrier between viewer and story. Yet it also raises interesting questions: do we want to be that close to something so disturbing? How does that affect our emotional well-being?

8. Horror's Ethical Questions

Horror occasionally faces criticism about its content. Some worry that graphic violence or extreme themes might desensitize viewers or glamorize harmful behavior. Others question whether some horror is exploitative, using real tragedies or painful topics just for shock. These ethical debates persist, though many fans argue that horror, like any genre, can include questionable material if not handled responsibly.

It is worth noting that horror can also raise empathy. While a shallow slasher might treat characters as disposable, a deeper horror film might highlight their suffering, prompting viewers to root for or sympathize with them. This nuance shows that horror's ethical impact depends on how it is created and perceived. In the end, personal thresholds vary. Viewers must decide for themselves which content aligns with their comfort and values.

9. The Lasting Power of Monsters and Myths

One might ask: why do we keep returning to werewolves, vampires, ghosts, or other old creatures? The answer lies in mythic resonance. These beings tap into beliefs older than modern civilization. They speak to the mysteries of nature, the afterlife, transformation, or curses passed through generations. Each era reworks these myths to reflect current themes. For instance, a vampire story might highlight loneliness in a city full of social media users who still feel isolated.

Such classic monsters have near-infinite replay value, as they embody broad ideas: the struggle between humanity and animal instincts, the danger of eternal life, or the regretful spirit trapped by past deeds. Writers add new layers—maybe the ghost lingers because of an unsolved injustice, or the werewolf's transformations result from a genetic experiment. As long as these core ideas remain meaningful, these archetypes will resurface in new ways.

10. A Global Tapestry of Fear

Horror is not confined to one culture. Every region has its own legends and beliefs about malevolent forces, shapeshifters, or restless spirits. With modern media, people can watch scary films from around the planet, broadening their understanding of what fear looks like elsewhere. A ghost from Southeast Asia might look and behave differently than one from Europe, but the underlying dread is similar. This cross-cultural sharing enriches the genre, leading to new hybrids.

Global connections also show that fear can reflect local history. A place affected by war may produce horrors about cursed battlefields, while a region dealing with urban sprawl might create tales of lost souls in abandoned buildings. This variety ensures that horror remains fresh, as each culture's storytelling traditions shape new nightmares. Audiences learn about the world while getting a scare, forging empathy through shared dread and curiosity.

11. Looking Ahead: Possible Directions

What might horror look like in the coming years? One possibility is deeper immersion via technology. VR has yet to reach its full potential, and as devices become more affordable, more viewers might want to step inside a haunted environment. Another direction is even greater focus on social horror, where the scariest elements come from problems like inequality, mass surveillance, or misinformation. Creators may continue exploring how collective fears shift when society faces abrupt changes.

We might also see more interactive stories that adapt to viewer choices. For instance, a streaming service could release a horror series where at key moments, the audience votes on plot twists in real time. This merges the community aspect with the tension of not knowing what path the story will take. While some might prefer a fixed narrative, those who like unpredictability could find it thrilling. Each wave of technological change opens new paths for horror to surprise us.

12. Balancing Tradition and Innovation

As horror moves into the future, it must balance tradition with invention. Part of horror's charm lies in familiar elements—jump scares, shadowy halls, haunted dolls, or suspicious neighbors. Fans rely on these tropes to know they are in the "comfort zone" of a scary experience. But overusing them can lead to predictability. Clever twists on standard tropes keep them lively. For instance, a haunted doll might be harmless while the real threat is something unexpected, flipping the audience's expectations.

Innovation also arises when creators question assumptions. Maybe the typical villain is not actually evil, or the setting is so ordinary that viewers do not suspect a hidden menace. A unique approach is especially valued in a genre known for repeated structures. Because horror welcomes experimentation, it can hold onto classic thrills while still providing fresh angles that engage new fans.

13. Personal Meaning and Memory

Horror's final impact is personal. Many recall the first time they were frightened by a certain film or bedtime story. That memory can shape how they see the genre: some vow never to watch it again, others become lifelong devotees. The best horror experiences remain etched in our minds. Years later, we recall the friend's house where we huddled under blankets, or the moment we almost screamed in a crowded theater. These stories pass among friends, or even across generations, becoming mini-legends of their own.

Because horror can be an intense genre, it leaves strong impressions. A well-crafted scene might continue to cause chills long after viewing. For some, this becomes a point of pride, showing they overcame the scare. For others, it marks a boundary they know not to cross. Either way, horror rarely leaves people indifferent. It either enthralls or repels, and in both cases, it makes an impact.

14. Horror's Role in Broader Entertainment

Though horror stands on its own, it also influences and is influenced by other entertainment fields. Action movies sometimes borrow intense chase or suspense scenes, while superhero stories include horror-themed villains or moments of darkness. Fantasy can merge with horror elements, creating grim worlds. Even musicals sometimes incorporate spooky tropes if the plot suits it. This cross-genre flow ensures that horror does not sit in isolation. It interacts with popular culture, shaping and reshaping it.

Meanwhile, comedic shows reference classic horror icons, comedic animations spoof slasher tropes, and detective series dabble in supernatural cases. The edges of horror blend into mainstream media, showing how it extends beyond direct fear to color different narratives. When creators want to add a dash of tension, they often look to horror's toolbox—shadows, suspenseful music, the reveal of something lurking behind a door. Thus, horror techniques are recognized as valuable across storytelling landscapes.

15. Personal Growth and Facing Fears

For some fans, engaging with horror leads to personal growth. They watch a film with disturbing scenes and learn to manage the anxiety. This can boost confidence in dealing with everyday stressors. Others use horror as a metaphor for personal issues. If they see a character conquer a demonic presence, they might feel hopeful about tackling their own challenges, whether that is speaking up at school or addressing emotional hurdles.

Of course, horror is not a substitute for actual therapy or life lessons. But it can prompt reflection. A person who sees a story about family curses might think about inherited habits or generational traumas in their own life. A zombie tale might make them consider how groups react under crisis. Thus, horror can lead viewers to examine personal or social topics in a safe, fictional framework.

---

16. The Legacy of Horror Creators

Throughout our exploration, we have mentioned directors, writers, and makeup artists who shaped the genre. Their impact goes beyond box office numbers. They inspire future storytellers, from ambitious students making short films to established creators wanting to add new layers to standard plots. The line of influence continues as each generation builds on what came before. A classic monster design might be updated by a modern artist, paying tribute while adding fresh details.

This legacy stretches back to the earliest silent horror works and forward to today's digital experiments. Each era has figures who left a mark, from authors like Mary Shelley or Bram Stoker to directors like Alfred Hitchcock, Tobe Hooper, or contemporary creators forging new territory. Their combined efforts form a mosaic that reveals horror's depth and adaptability, ensuring it never stands still in a single style or idea.

---

17. Caution for All

While horror can be thrilling, it is not for everyone. Some viewers are deeply disturbed by gruesome images or intense psychological tension. Others find certain stories conflict with personal beliefs or bring up painful experiences. It is wise for people to know their limits. Just as some folks love roller coasters while others avoid them, the same applies to horror. Listening to one's feelings and stepping away if something becomes too much is important.

Moreover, creators must acknowledge the weight they hold. A film or book can leave a lasting impression. Using shock purely for shock's sake or focusing on extreme violence without depth can alienate many. That does not mean horror should be censored, but it does mean that creators who approach challenging topics should handle them with skill to avoid trivializing real trauma. In the end, horror's power lies in making us feel deeply—so it helps when those feelings are crafted with care.

18. Summing Up the Book's Journey

We have examined horror from various angles. We started by defining what makes a horror movie and why people are drawn to fear. We discussed the brain's reaction, older traditions, monsters, how horror grew in cinema, the directors who shaped it, and the many themes that exist. We also looked at how horror appears worldwide, how music and special effects affect it, moral dimensions, the fans who adore it, fear in different cultures, horror on small screens, and even low-budget projects. We considered younger viewers and modern trends. Finally, we have arrived here, seeing how horror remains a vibrant phenomenon.

These chapters show that horror is not a narrow field. It is a sprawling canvas where creators and audiences interact, each contributing to a cycle of imagination, shared emotion, and ongoing renewal. It addresses our darkest corners while offering hope that we can handle them, or at least share them with others. Sometimes it warns us. Sometimes it helps us laugh at our jitters. But it always leaves a mark, forging memories that last and stories we retell.

19. Diversity in Fear

One final point: horror is as diverse as the people who engage with it. Different languages, settings, and styles produce a vast range of unsettling tales. Some focus on ghosts with cultural roots, others highlight modern technology running amok, and still others revolve around plain human cruelty that can be more chilling than any supernatural entity. This diversity is a strength, ensuring that no matter someone's background or taste, they can find a subgenre that resonates.

We see collaborations across borders, with directors from different places teaming up, or streaming services promoting foreign horror to global viewers. Fans can watch a chilling short from one continent and a big-budget ghost film from another in the same evening. This cross-pollination fosters empathy by showing that fear is universal, even when expressed in unique ways. Each local myth or practice becomes part of a broader tapestry of spookiness.

20. The Ever-Open Door

Horror is often associated with doors: creaking ones that lead to basements, locked ones hiding secrets, or mysterious ones in abandoned mansions. These doors symbolize entrance into the unknown. At the end of this book, it is worth noting that the door to horror is never truly shut. There is always another hidden corridor, another shadow waiting to be revealed. For newcomers, horror can be an uncharted realm, a step away from routine. For long-term fans, it is a place they revisit, seeking the next surprise.

The genre's future likely holds more cross-genre blends, more inclusive voices, more advanced ways to immerse ourselves in fear. But it will also keep honoring the core spark that made it compelling in the first place: our shared curiosity about what lies beyond normal life, and our urge to feel fear and then conquer it. As long as we remain human, with hearts that race and minds that imagine monsters at the edge of sight, horror will stay with us.

Concluding Note

From haunted myths to digital nightmares, horror persists because it is woven into our emotional DNA. We dread the dark yet gather to swap chilling tales, half hoping for a scare. This push and pull between fear and fascination will continue to guide horror's path. By understanding its history, psychology, cultural ties, and modern directions, we see that horror does more than cause jumps and screams. It connects us across time and place, reminding us that in the face of the unknown, we are neither alone nor helpless. We hold each other's hands, take a breath, and open that creaking door—ready for whatever lurks in the gloom, because sometimes a good scare makes us feel more alive than ever.